Landesmuseum Joanneum

Shiny Shapes

Arms and Armor from the Zeughaus of Graz

Edition Joanneum

Thomas Höft – Text

Alexander Kada – Book Design and Photo Concept

Angelo Kaunat – Photography and Lighting

SpringerWienNewYork

The collections of a museum are a reservoir, constantly revealing something new when examined from various viewpoints.

The Landeszeughaus in Graz, with its unique collection of historical arms and armor dating from the period of the Turkish Wars, provided the subject for the first of a series of books by Springer (the next will be dedicated to Eggenberg Castle) aimed at taking a new look at the collections and buildings of the Landesmuseum Joanneum. Our intention here was to bring the object down from its "pedestal", make it approach the visitor, tell its story, convey emotion.

To support us in this project we drew on external observations, making them an integral part of the museum's identity.

Thomas Höft, Alexander Kada and Angelo Kaunat - author, designer and photographer - cooperated closely in Graz as the new concept evolved. Their work endeavors to present the museum pieces as contemporary, to be understood as lively objects in an immediate context; helmets become faces, armor becomes clothing. Suddenly one is aware of the obvious connection between formal gesture and emotional function. In the16th century the court amorer Michel Witz was commissioned to design a helmet both fierce and elaborate. The intention of the armorer/designer should be made clearly recognizable, the meaning adequately presented. This explains the uniqueness of the following volume: through the authors' eyes the reader experiences a fascinating documentary, newly staged by eliminating pretentious gestures and telling the story of the objects in terms of our present-day awareness and technology.

Formal research is the basis for responsibly conveying a visual history that vividly outlines the social structures during the period of the Turkish Wars, and is also capable of telling the other side of the story.

A respectful and intensive discussion of the life and character of the objects today can provide rich information regarding the representative function, fashion, hierarchy, status and purpose of these iron personalities.

An aesthetic approach brings the pieces to life; they become actors enticingly "beaming" the reader into bygone ages; they appear fast and dynamic, sad and proud, fragile and brutal, at times even grotesque. As such, the emotional impact of the book goes beyond mere historical information, bringing alive a chapter of our province's history.

My thanks go to the authors for their impressive work that has set new standards for all future publications of the Landesmuseum Joanneum, and to Springer for their support.

May our readers enjoy a fascinating journey through the past.

Barbara Kaiser
Director of Landesmuseum Joanneum

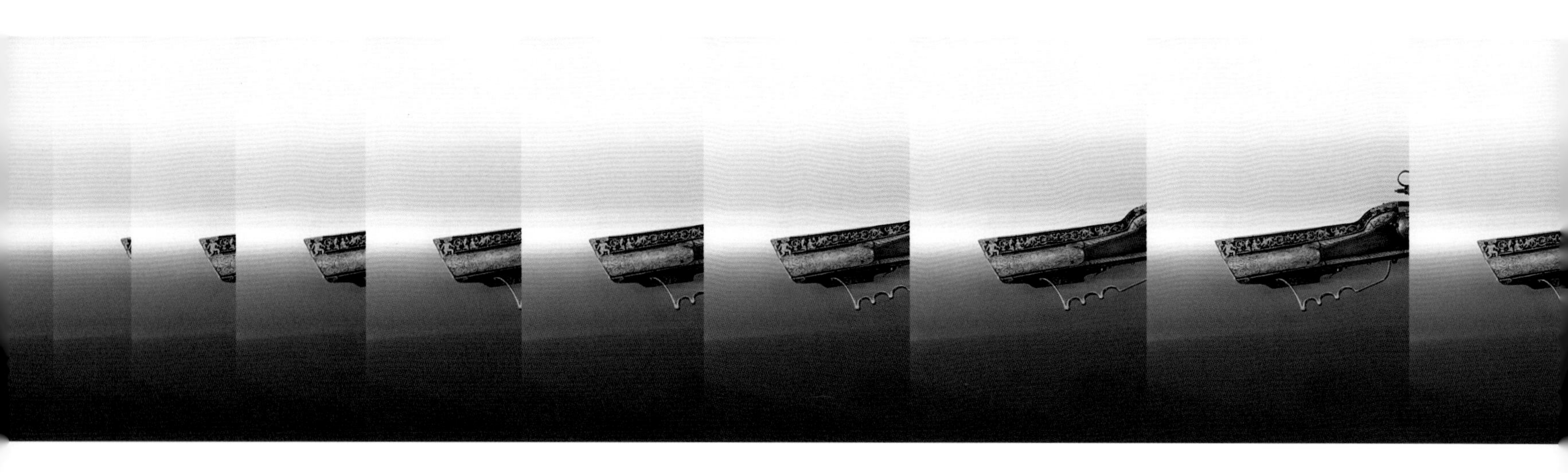

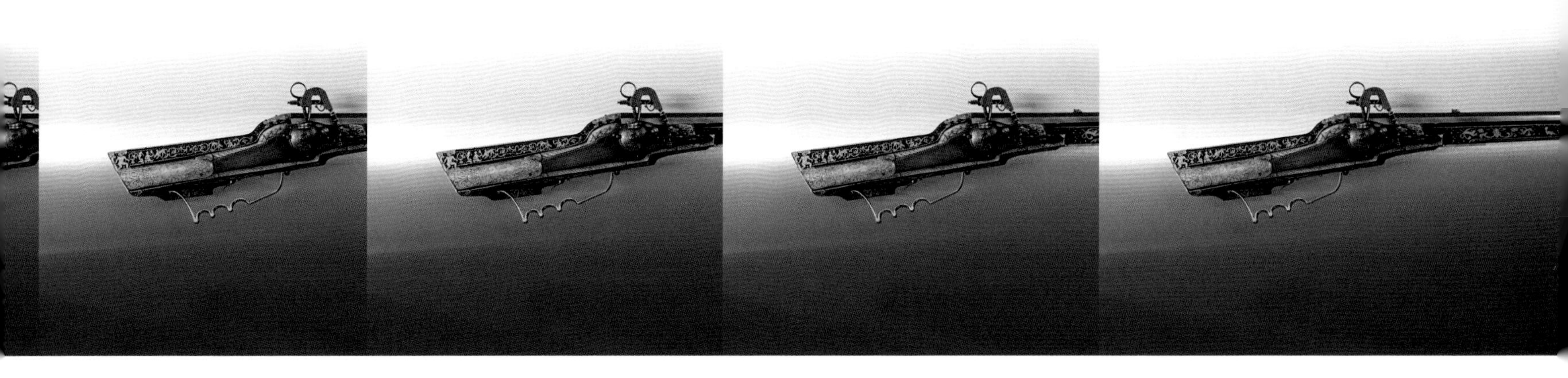

Prologue

Guarded Places

How harmless things appear when viewed from a distance.

In the middle of downtown Graz, on Herrengasse, one can sit at a sidewalk café and take a look at history. The road is closed to traffic, streetcars dictate the rhythm of the street and the panorama of house facades looks suspiciously like a travel guide. The historical Rathaus shows off with exaggerated designs of the late Renaissance, grotesque figures, industrially reproduced patterns of an occupied past which, in turn, have become history: "Gründerzeit" pomp, and for the moment, perhaps even a new-found identity.

Adjacent is the Landhaus. This time pure Renaissance, built in 1565 by the Italian, Domenico dell'Allio, as seat of the Styrian estates, representatives of local aristocracy and clergy. Right next door is a tall, narrow building, rather plain with huge closed iron doors: the Landeszeughaus. This armory for the Styrian estates was designed and built between 1642 and 1645 by the Swiss master builder Anton Solar. Its portal is flanked by two statues, theatrically dressed figures in war pose: Mars and Minerva, the ancient gods of war.

The presence of all too much history can also result in distance. It is no problem to enter the armory. The streetcar conductor cautiously rings his bell warning pedestrians who want to cross the street in front of him. And instead of a special pass, one only needs an entrance ticket for the Styrian Landesmuseum Joanneum to which the armory now belongs. However, once within the dim, narrow walls of the arsenal's five floors, the picture changes. One is met by row after row of wooden stands clad in armor - lances, spears, swords, firearms - 280 years old and older - thousands of objects, hanging from the ceiling, covering the walls, filling the room, overwhelming the viewer. A sequence in iron, a sober drama of the masses, the pure functionalism of power. The building is unique. Apart from a single baroque armory in Solothurn, Switzerland, all others have been scattered, destroyed. Slowly one begins to notice detail, the galleries of armor are perceived as individual pieces, quality emerges from the masses. Handicraft, not industrial mass production, though masses had to be outfitted - against the Turks, against the insurgents from Hungary. And thus one has arrived in the capital of a borderland, a mighty bastion on the edge of the western world.

The Forbidden Zone

Military weapons are state secrets.

Today the type, quantity and storage of weapons is a subject of utmost secrecy just as it was 500 years ago. The arsenal is a guarded place giving birth to speculations, rumors and interpretations which must necessarily remain unconfirmed. The reasons for this are mainly of tactical nature. A possible enemy is never sure how one plans to defend oneself. And one's own army is never fully aware of the true possibilities, as well as limitations, of self-defense or aggression. They must rely on trust. Thus, the visible exterior of the arsenal is ideally a symbol of trust. It is a bulwark whose protected interior conceals the means of one's own strength.

When Anton Salomon Schmidtmayr from Varaždin arrived in Graz in May, 1661, to view the artillery which had been promised him for border defense, he was refused entrance to the arsenal. Schmidtmayr was armskeeper on the military frontier supplied by Graz, thus a friend on the front line, exposed to acute danger. Nonetheless, the armskeeper in Graz, Sigmund von Klaffenau, was obliged to bring the artillery into the courtyard. The doors of the armory remained closed even to his colleague.

Weapons are state secrets. The Zeughaus in Graz is a guarded place, charged with symbolism. Yet due to a combination of lucky circumstances and special efforts, caution and cunning it remains intact today. It offers the rare opportunity to break through protective barriers, enter the guarded interior and view the contents of symbolic significance. A walk through the Zeughaus of Graz is not a normal museum tour but a trip through time into a once forbidden zone which, even today, contains all of the menace and threatening energy inherent to the building's character - for those who know how to interpret its history.

In 1748 when Empress Maria Theresia resolved to give up the armory - the danger of a Turkish invasion was over and the obsolete arms were considered not worth more than scrap metal - the Styrian estates protested. The armory was to be preserved as a memorial to the history of their country and the valor of their forefathers. Maria Theresia agreed and the symbolic character of the Zeughaus was established.

The Exterior

Mars and Minerva, the figures flanking the Zeughaus portal, stand for two aspects of war. Mars, the fiery god of war, embodies in the period's style, the fury of battle - all the courage and strength necessary to triumph. With his beard and head of curls, he is the personification of high-ranking 17th century military figures. This Mars is contemporary and present-day, only the antique costume, sword and shield identify him as a Roman god.

ANNENHOF
DENN SIE
WISSEN NICHT,
WAS SIE SEHEN

Minerva represents the opposite. If Mars is the anger of war, she is the strategy. Planning, reason and order - her Greek counterpart, Pallas Athene, sprang from the head of Zeus - thus cerebral from the beginning. No wild slaughter but the artful planning of the campaign. Her battle is to protect the homeland. Stylized in antique costume, she stands in front of the Zeughaus, contemporary like her partner.

Together these figures are tangible representatives of the Zeughaus creed of art. Battle and defense, strength and strategy are incorporated here. The building needs no further decoration. It is sober, functional. The only exception is the large portal with its Styrian panther above the door as symbol of the province, and the five crests of the estates of 1644 on the door beam - paying honor to those elected representatives who ruled the land. The first step toward construction of the Zeughaus was taken in 1639 with the purchase of the Rattmannsdorf House, located next to the Landhaus. It cost 4000 guilders. In comparison, 80 liters of grain cost one and half guilders at the time. The newly purchased house was torn down beginning on the rear side. At the same time building material was brought in so that new construction could begin in 1642. In 1644 the front wall of the old house was razed and in the same year the building was completed. Total cost was approximately 14,000 guilders. Construction was led by Anton Solar whose contract with the Styrian estates included exact measurements (52.5 meters long, 11.5 meters wide) as well as extensive guarantees. The contractor was responsible for repairing, at his own expense, any shortcomings which appeared within the first year after completion.

The ground floor, used to store heavy artillery, canons, munition and wagons, had four big doors on one side which opened to the courtyard, thus enabling rapid access to equipment. At the same time this courtyard was separated and protected from the Landhaus by an office tract. The arcades so characteristic of the Landhaus today are actually a historical addition built at the end of the 19th century upon removal of this office tract. Thus the cheerful openess and Italian atmosphere of the Landhaus courtyard, the manneristic ornate arcades which lend the Zeughaus facade a romantic air are a product of the "Gründerzeit".

The once hermetically sealed exterior is today but an echo of the past.

For historical information about the Landeszeughaus in Graz:
Peter Krenn, Zur Geschichte des Steiermärkischen Landeszeughauses in Graz;
in: Festschrift 150 Jahre Joanneum 1811-1961; Graz 1969.
New findings regarding the building's construction:
Leopold Toifl, Spurensuche im Zeughaus;
in: Jahresbericht 1995 Landesmuseum Joanneum; Graz 1996,
which was also the source of the story about Salomon Schmidtmayr.
The question of money value and war financing is discussed in:
Othmar Pickl, Kriegsfinanzierung als Faktor der Wirtschaftsbelebung:
in: Geschichtsforschung in Graz; Graz 1990.

Discovering the "Other"

January, 1998

Walked with a group of German tourists through the Zeughaus.
Heard their conversation, saw surprise written on their faces,
as they danced through the rows of arms, giggling.
Entertainment in the Land of the Knights.
Didn't follow as they, tired after three minutes,
looking for something new
climbed the stairs,
only to find more of the same.
On the second floor stopped by the canons.
Wanted to stick my head inside a barrel...
maybe a new insight.
Observed the "Hakenbüchsen",
trying to imagine the kickback.
Believing it could catapult one back into another age
where the iron is not,
vaseline-shiny with public care,
only a vestige.

Wheel-lock Guns

"The worst is the rust,"
says the curator,
who discourages my touching.
"Like a disease - invisible from the outside - eating through the metal.
And then, from one day to the next,
spreading like cancer.
You must know,
armor is a second skin.
Metal breathes.
It has pores."

God's Plagues

A partially preserved fresco on the south side of the Graz cathedral compares the disastrous Turkish invasions in August, 1480, with the effects of the plagues of locusts and disease already ravaging the land.

Years of terror captured in a religious composition, as if these occurrences were God's punishment to mankind. And yet still hope of mercy and compassion - depicted by the Trinity which thrones above the scene of horror - a hope which grew more fervent as the conditions worsened.

At the end of the 15th century, an extended warm weather period in central Europe brought on catastrophic consequences. As early as March, the trees in Styria were in blossom. Lack of precipitation and scorching heat soon made water a scarcity. In 1473 so many wells went dry that water had to be brought in from Carinthia. The crops withered and swarms of locusts pushed far into the north. One eye witness, Ursula Silberberger, abbess of the monastery in Göß, recorded the following in her chronicle:

"On August 17, in the year of 1478, a huge number of locusts suddenly appeared and swarmed through the entire area. The air was so thick, as if it were snowing, and the sky was obscured. These locusts were as big as titmice or siskin (gold finches), and in many places they destroyed entire crops down to the roots. There were so many that when they landed,

Flint-lock Pistols

birch trees and hazelnut bushes bent under their weight. They lay so thick on the ground that one sank in to the calves and had to wade through them as if through snow or sand." At the time it remained warm and the locusts survived way into winter nourishing themselves on the winter seed.

However, this insect swarm of biblical dimensions was not the only thing responsible for the apocalyptic judgement day-mood. In 1347 an epidemic broke out in Genoa, probably imported by ship from the Crimea, which in the following years eradicated a third of Europe's population. After the first devastating appearance of the pest, it became endemic. New seats of disease continued to break out repeatedly and for more than 300 years the pest remained an uncontrollable and deadly threat whose means of transmission were unknown. The assumption that such inexplicable and ungovernable phenomena were signs of God's wrath was widespread, especially since Catholicism of the day considered both punishment and mercy as signs of God's reaction to human behavior. Religious fanatism, sectarianism and repentance movements were widespread. The search for those considered guilty claimed victims.

Much has been written as to how social systems deal with inexplicable threats and what criteria those responsible use to define and seek out the enemy. The fact that victims are, above all, individuals standing outside of the ruling conceptual and religious system is perfidious reality. Jews, branded as "thorns in the flesh of Christ" and accused of poisoning wells, were ideal targets for the projection of obscure fears. First pogroms and mass expulsions were the bloody consequence. The persecution of heretics and "witches" also appears as a desperate attempt to gain power over uncontrollable conditions. However, time and again efforts were also made to identify guilt and corruption at the heart of the governing system. The movement led by charismatic religious reformers such as John Huss found fault not in the "others", the foreigners, but in the outlived, dishonest and rigidly egoistical position of the Church which had grown away from its original roots. Just how strong this fundamental criticism

Mortar 17th century

was to become is demonstrated by the Lutheran reform movement. Not that the customary victims were completely spared. In written attacks, Luther proposes driving all Jews, who refuse to convert, into a city and "igniting the four corners of the town". The social and intellectual consequences of Protestantism will be discussed later. Meanwhile the year 1480 brought along another plague which quickly developed into an existential threat: for the first time Moslem soldiers directly attacked central Europe. And at first, nothing seemed able to stop them.

On August 6, 1480, approximately 16,000 Turks reached Styria after passing through Carinthia. They pitched camp on the Aichfeld near Judenburg and within eight days swept through the land in three groups, murdering and plundering along the way. The swift riders spread to the Lavant Valley, to Rottenmann, Göß, Leoben, as far as Bruck, and also did not spare the area around Graz. Finally they withdrew through Radkersburg as quickly as they had arrived.

February, 1998
In the Zeughaus searched for traces of the Turks,
For what the OTHERS were,
Nothing there.
No shield, no spear, no helm.
The other does not exist in one's own inner realm.
He is warded off,
The view of him shut out
like closed visors on the iron hoods.

The Turks

It is impossible to describe the Turks from the standpoint of the Zeughaus. Nonetheless we do know what the picture of them must have been. What caused the fear, what caused the menacing threat. This sketch of

Jack 17th/18th century

the opponent is biased, it is an interpreted view. And certainly the Turks had their own, complementary picture. For an objective analysis, it is important to discard the moralizing handicaps which both sides possessed and expose the function which this moralizing served. The war was exceptionally brutal and gruesome. It was a matter of dominance, power, conquest. It was also a matter of the priority (and preservation) of a way of living, of culture and social concepts, and of religion. Every judgement of others reflects one's own code of values and acts as both justification and propaganda.

As opposed to Christianity, Islam does not include a missionary mandate as part of its written doctrine. Although Mohammed's teachings should be spread throughout the world, forced conversions such as those carried out with regularity by Christian conquerors are hard to reconcile with Islamic ideas and have political rather than religious foundations. A distinction is also made between "heathens" and those possessing a "holy book". Islam is a prophetic religion and considers its religious doctrine the ultimate consummation of those beliefs introduced by Judism and Christianity. Mohammed bade his followers to be tolerant of monotheistic believers who have a holy book. This tolerance derives from the genealogical origins of Islam.

The book of Genesis tells the story of Israel's ancestor, Abraham. This man appointed by God, is unable to have children with his own wife, Sarah. Then God bids him to take the Egyptian maid, Hagar, into his home and she bares him a son, Ishmael. God establishes a covenant with Abraham and institutes circumcision of the men of this folk as token of the covenant. And as confirmation of the elevation of Abraham to "father of the peoples" his wife Sarah finally becomes pregnant as well. In her jealousy she drives Hagar from their home. However, God consoles the despairing mother, Hagar, and prophesizes that her son will also give rise to a great nation. The Arabian Mohammedans consider themselves the descendants of Hagar and Ishmael consequently explaining their claim as God's chosen people, as well as their ambivalent relationship to Judism and Christianity. This remains so to present-day. The bloody conflicts regarding the "holy cities" in today's Israel are only comprehendable in this context: all three world religions consider these cites holy for a different reason. And if one follows the actual discussion between Jews

Jack 17th/18th century

and Palestinians, the humiliation of Hagar and Ishmael by Sarah is still the best image to describe the strained relations between members of these religious groups.
Mohammed's appeal for tolerance is nonetheless limited by the rather broadly defined "defense of faith" principle which allows no defamation or degradation of Islam. And of course, by political interests. The expansion of Islamic influence - not just the religion itself - by means of force is not only accepted but demanded. This is based on the early Islamic concept of a world divided into two houses: the "House of Islam", governed by Moslem rulers, and the rest of the world or "House of War" which must be conquered. Those who subject themselves voluntarily are met with the same tolerance accorded to those possessing a "holy book". This, however, does not mean living side by side as equals, but the total submission of the adversary and expansion of the Islamic culture.

As opposed to the New Testament, the Koran and Sunna – recounting the words and deeds of the Prophet – provide numerous clearly defined norms and rules of behavior. For the most part they are a practical code and, though subject to interpretation, take a clear position on daily life thus resembling the Commandments of the Old Testament, compulsory for all orthodox Jews. The right to plunder during war and then distribute the booty among the warriors is part of Islamic decree and provided a concrete incentive for military conquest. Many Islamic soldiers actually earned their primary means of living from war spoils. Payment of reparations by the defeated party is also clearly defined and many sources report an easing of taxes after Islamic leaders replaced Christian authority.

Of course, Islam - like Christianity - is not the monolithic block which it propagates in order to intimidate the opposition. Even at the height of its power, severe internal conflict characterized the Islamic Empire. Turkish hegemony was earned only after numerous set-backs with tremendous losses. The legendary victory around 1400 of the Mongols under Tamerlan over the Turks recalls how the Turkish leader, Bajazet, was then displayed in a golden cage.

Original home of the Turks is Central Asia. From the 11th century aggressive Seljuk leaders rapidly conquered large parts of Asia Minor. Torn by internal conflict and divided into small rival groups, it was finally Sultan Osman who, after 1300, managed to consolidate his position of power and embark on expansion campaigns. From their capital Adrianopel, the Ottoman Turks conquered the Serbs (1389 in Amsfeld), crushed crusader troops at Nikopolis in 1396, and then conquered Wallachia. The conquest of Constantinople in 1453 by Mohammed II, the Conqueror, confirmed for once and for all the Ottoman position as a world power and dangerous threat to the West. After the fall of Bosnia, the way to the heartland of the Habsburgs lay open.

Upheaval in the Society

The Plague of God fresco on the cathedral of Graz can be viewed as a kind of late-day "apotropaion" - a charm to ward off evil. Placed on the outside wall, it is both warning and attempt to frighten opponents in a period of desperate upheaval. The assaults from outside seem to reflect (or act as punishment for) the desolate internal situation of the Empire. The political and social conditions at the time of the first Turkish onslaught are, to say the least, confusing ; contemporaries of the day often referred to things as chaotic. Traditional positions of power in the medieval society had long been falling apart. A schism in the Papacy (at one time, three popes reigned against each other) expressed the basic weakening of central religious power. And even the prestigious Emperor of the Holy Roman Empire of the German Nation (a title which became popular in the 15th century) was just a vulnerable monarch subject to attack.

Local authorities - princes and ecclesiastical rulers - often managed to establish strong economic positions. Flourishing trade, strengthening cities and the increasing desire for civil liberties - all of these developments were too complex for the simple feudal order. Society was in upheaval, the familiar systems of knowledge and faith gave way to new structures based on measurement and numbers. In fact, in the century before the first massive onslaught of the Turks, regional structures were set up to enable a correct census of the population for the first time since the Roman Empire. Population counts - as basis for property evaluation and taxation - were reintroduced and perfected. On the one hand was a new objective relationship between ruler and subject; on the other hand was mistrust, disloyalty and the collapse of the old social framework.

It was the age of upstarts, war profiteers and bands of roaming mercenaries; local lords regard the weakness of the central authority as an opportunity to improve their own situations. The wheel of fortune spins quickly and victims lie on all sides. At first the Turks present only a minor part of the danger. Often it is the local rulers themselves who bring in the murdering plunderers who terrorize the rural population. Yet at the same time, the Plague of God fresco marks the beginning of change. The consolidation of relations, the strengthening of structures also falls in the period of the first Turkish wars; and perhaps it was this completely "foreign" threat which finally led to the inner - though somewhat fragile - stability of the Empire.

This stability is linked with the increasing power of the Habsburgs under the rule of Emperor Frederick III (1415-1493), a man who paradoxically earned little praise for his political skills - even in his own lifetime. If politics are to be understood through political symbols, then indeed his reign was characterized by a series of devastating set-backs. Threatened by his own brother, repeatedly besieged and defeated, he became fearful and false, even breaking his word - truly no radiant figure. Yet in practical politics Frederick emerged as a winner for the simple fact that he survived. By hesitating and evading, dodging and delaying he managed to save his own life, and his son guaranteed the family existence.

As for the children: from today's point of view the vehemence with which the presence of sons, of male heirs, was regarded as a factor of stability seems almost archaic. Territorial security and the guaranty of continuity were directly linked - for better or for worse - with the ruler's offspring.

The authority over a child also plays a role at the beginning of Frederick's rise in power. Albert V, King of Hungary and Bohemia, died on a campaign against the Turks in 1439 - (death at battle of direct heirs to western powers was responsible for the first decisive changes caused by the Ottomans in the Occident). Albert's son, Ladislaus Postumus, heir to the title, becomes a pawn in the hands of those rivaling to take over the crown. Frederick names himself guardian and wants to reign in his name, however, the estates of these varied and disunited lands revolt: a dangerous power struggle between emperor and regional nobility. Finally Frederick III is forced to return Ladislaus to his enemies. Shortly afterward, the child dies, probably poisoned. Thus George Podiebrad becomes King of Bohemia and Matthias Corvinus, King of Hungary - both elected by the powerful estates of their countries. Even in his own realm the Emperor must struggle to survive. His brother, Albrecht, claims part of the regency and is supported by large parts of the Austrian estates. ("Go back to Graz," cry the Viennese - for according to the Emperor's line of inheritance he is Duke of Styria and Carinthia.) By negotiating and yielding, though not completely withdrawing, Frederick manages to deal with his brother, whose death in 1463 finally eliminates him as a rival.

The Hungarian Wars

The greatest danger for Emperor Frederick, however, is the mighty King of Hungary, Matthias Corvinus. Elected by the majority of Hungarian estates - only a minority backed the Emperor - Corvinus is also capable of defeating Frederick militarily. It was the attacks of Hungarian mercenaries - even before the arrival of the Turks - which prompted Styria to reorganize its provincial defense system. Though prepared for action, these troops are no match for concentrated Hungarian attacks. Military conflict continues intermittantly from 1446 to 1490. Again and again the Emperor is defeated though never completely destroyed. Matthias Corvinus is also unable to change the most important condition: when the Corvinian dynasty dies out, Hungary (known as the Land of St. Stephan's Crown) will return to the House of Habsburg. Upon his death in 1490, Matthias Corvinus leaves no male successor - only an illegitimate son who he did not succeed in making his heir. Thus Emperor Frederick III survives another rival. However, the price is high - over forty years of plundering and devastation - not only from the Hungarians, but also from mercenaries which the Emperor had brought in and then was unable to pay. These men recoup their losses on the population. Contempt for the Emperor increases. Finally Frederick's chronic financial problems cause a wave of inflation directly induced by the Emperor's minting of inferior coins, "Schinderlinge" - which are soon rejected as payment. One of the worst consequences of the dramatic situation: Jewish pogroms carried out with a brutality unknown till now in Austria and the ridicule of Frederick as "Emperor of the Jews". The crisis is thus not only social, but above all economic. And the search for those to blame functions in the same well-known manner ...

The Hungarians prove to be no less brutal than the Turks. Descriptions of slayings and arson attacks cast light on the living conditions in an area hopelessly caught in the chaos of changing alliances. When Matthias Corvinus orders the attack of Radkersburg and Fürstenfeld in 1480, Radkersburg is first leveled by massive artillery while Fürstenfeld burns under a shower of flaming bolts. Inhabitants who survive the fire are hauled off. Finally only three houses and the monastery remain standing. The latter is completely ransacked, one monk assassinated and the other eleven held captive for 15 days and "treated like animals," according to a later account by Prior Augustine. This eye-witness further reported that "degenerate Franciscan monks" from Hungary participated in all these despicable crimes.

Borders become blurred. Who is friend, who is foe? Sometimes this distinction can only be made too late. Citizens of small, defenceless towns as well as peasants become the very first victims. And fear of the Turks is an instrument shamelessly exploited by all parties - even before they appear at the gates. Thus Matthias Corvinus cleverly presents himself as hero in the face of the Turks. Using them as an excuse to place all fortifications under his protection, he then takes over the castles of Archbishop Bernard von Rohr of Salzburg and of Bishop Christoff von Trauttmannsdorff of Seckau. Both of these men are only too happy to cooperate with the Hungarian as the Emperor had previously tried to replace both of them with his own personal confidants.

Corvinus vigorously stirs public discontent against the Emperor - though he himself takes a seemingly subservient position in personal confrontations (political symbolism). The Hungarian also supports Andreas Baumkircher, a nobleman from Carniola, in his feud with the Emperor. A feud is an extreme course of action including the use of force in cases of serious transgression, taken by people of rank according to an exact set of rules - also admissable in disputes with the Emperor. Baumkircher was a loyal subject of the monarch, and had lent huge sums of money to finance the Emperor's mercenaries. However, when all hope of repayment disappeared, Baumkircher wrested his way to the head of Styria's aristocratic opposition, and waged war against Frederick, a further occasion for looting and devastation which weakened the land. This conflict ended in the Emperor's breach of promise. Assured of safe conduct, Baumkircher and leaders of the noble opposition were ordered to Graz where they were then decapitated. Strangely enough, in the negotiations between the Emperor, Baumkircher and Corvinus, the Styrian estates agreed to pay Baumkircher's mercenaries who were still roving the land, thus finally forcing their withdrawal. This shows that not only the Emperor and the Hungarian opposition swayed in their loyalties, but the Styrian estates as well. What strength could the nobility afford? When was it opportune to agree with whom? Frederick's unscrupulous and violent intervention in the Baumkircher feud seemed to incite - if not respect - at least fear in his opponents. This small melancholic man, with his strong aversion to the gluttonous feasts and drinking bouts common to the ruling class of the day, was considered eccentric purely on account of his unsociability. Nonetheless he managed through unpredictable behavior and clever tactics to secure political power for his family and for his son Maximilian, the "last knight", whose talent brought new hope even to the Emperor's critics. Maximilian succeeded in further consolidating the Habsburg's claim to the Hungarian crown. His father's goal was achieved, Hungary defeated. The more dangerous foe, however, was about to enter the scene.

Defense Regulations

Emperor Frederick III cultivated the habit of sleeping long hours. His secretary, the humanitarian and later Pope, Silvius Aeneas Piccolomini wrote detailed descriptions about this as well as many other characteristics and peculiarities of the Emperor including his preference for vegetarian fare and his shyness towards, even aversion to women. The image of the sleeping emperor grew from a personal anecdote to a political symbol. It was typical that the Emperor usually refused a staging of his person through external, material images preferring to choose his own individual style - and exactly this fact often worked to his disadvantage. For instance, his renouncement of opulent clothing was not regarded as modesty but rather as stinginess. When he did decide on symbolic presentation it came to an éclat: the precious jewel-covered cloak which he wore for the Imperial Coronation in Rome filled the crowd with jealousy. They wanted to tear it to pieces (as was done according to old tradition) and in the ensuing tumult people were killed. However, just these precious gems aroused the Emperor's interest, not so much for their material value as for their power as magical objects in the tradition of alchemy. His high regard for mysticism, for hidden symbolic and energetic-philosophic correlations is reflected in his universalist symbol AEIOU which he often used as insignia. Thus the imperial images were of an inner-philosophical rather than of an authoritative-political nature. This view is verified by the chronicler Johannes Grünpeck who reports that though Emperor Frederick was very small, he had practiced from an early age the art of appearing regal; his facial expression was earnest and reflected serene modesty. Frederick used his own individual and very personal body language. The sleeping Emperor: for his opponents, a symbol of insufficient protection, of being left on their own, and of the lack of imperial solicitude traditionally expected of the emperor.

In 1470 a pamphlet appears, first in Vienna, then shortly afterwards in Graz, criticizing the Emperor [the so-called "Minoritenpredigt" (Minorite Sermon)]. The Emperor sleeps, ignoring the needs of his people. Did he not receive a sword at his coronation - not a book (the book in this case symbolizes unworldly, melancholic, scholarly circles), a sword to protect his lands? Yet the dangers of Turks and feuds do not interest him. He has no money? Now, do not torment the poor with new taxes, does he not have access to the property of the Church as if it were his own? Where is the income from the last decades? His subjects are not loyal? Yet this disloyalty results from lack of welfare and security. Money declines in value due to greed; the princes ignore the rights of their subjects and what happens? All remain silent, even the prelates and Pope. That is why an insignificant Minorite has to point out all these injustices. One has to take charge of things - to take care of oneself.

Take care of oneself: a ludicrous idea in a social system where the dependent position of the weak is only justified by the ruler's promise of protection. If this promise is no longer valid then the whole system of subordination begins to sway. The vassals demand more rights since they have been left alone to defend themselves. The Emperor is aware of the explosive nature of this development but cannot suppress the movement - for now, defense is a matter of survival. In 1443 the Styrian estates order the first general military conscription to defend themselves against the Hungarians. It is based on census structures that had already been implemented in the previous decades. Military enrollment was based on the parish, several grouped together formed a district.

The entire country was divided into 22 such districts, headed by 75 captains, all members of the nobility. Of the male peasant population, every tenth man was drafted (according to his physical constitution) as footsoldier, the other nine were responsible for providing food and equipment, as well as for carrying out the draftee's work at home during his period of military service. Twenty draftees formed a troop; specific instructions defined the type of arms to be used and the assembly points. Here they were joined by cavalry and mercenaries who were provided by the nobility, clergy, cities and communities. A landlord's responsibility for providing soldiers was based on the amount of tax he paid.

Tax payment as measure for participation in public armament: the symbolic meaning for the new upper class could not be clearer; responsibility brings new empowerment , a fact which Emperor Frederick was very reluctant to accept. Since the call to military defense was now controlled by the estates, Frederick was forced to make requests. The Emperor as petitioner was not exactly a role he fancied for himself, however he was economically dependent, needing more and more support as his own financial means declined. The Styrian estates quickly took advantage of their new authority to improve their situation. In 1462 they assembled in Leibnitz and agreed upon a new defense system coupled with a more efficient means of taxation. The country was divided into four tax districts, each with two tax collectors, to facilitate the rapid and effective collection of war taxes. The leadership of the provincial military forces was also stipulated. A general, appointed and paid by the assembly, had not only command of the indigenous forces but also the right to use tax money for hiring additional mercenaries. The estates - when unified - were a mighty rival to the Emperor in an area where he had shown obvious weakness.

Conscription

Though Emperor Frederick protested that a conspiracy was being plotted against him when the estates took over control of the military authority, he was too dependent on help to take action. The division of Styria into tax districts was then followed by exact specifications with regard to the number and kinds of weapons and military equipment to be held in readiness. In addition, towns and villages were ordered to set up munition depots and ration stores - the idea of an armory was born - and as such, a symbol of the authority of the Styrian estates.

Events in the year 1469 show how important individual responsibility for territorial defense can be for survival. Members of the peasantry in Upper Styria who felt immediately threatened made ultimate demands that their landlords takes step for regional defense. The peasants were rightly concerned about the growing Turkish threat as well as the spreading Baumkircher feud. Once again a menacing threat revolutionizes the traditional balance of strengths and power: if the ruling authorities remain inactive, then the commoners will take action themselves. This meant lifting the strict laws which forbade commoners to carry weapons - with all the ensuing consequences. The cities and nobility were alarmed. In cooperation with the peasants they agreed on a system of defense which completely excluded the Emperor. Noteworthy is the fact that this defense system is a collection of local precautions for the protection of manors and villages, developed to meet the needs of those in immediate danger. The defense plan is based on the parish, small zones from where the protection is organized. A local captain was responsible for a warning system; bell-ringing and fire signals were devised, and every parish was responsible for securing a certain sector of the border. Supervision of these areas rotated once a week and was organized by landlords and representatives of towns and villages. A cooperative concept which strengthened the idea of personal responsibility thus leading to the development of a new self-confidence so far unknown to customary life styles. A self-confidence which, especially for the lower-classes, was by no means a matter of course and, in fact, very modern.

Just one year later, the new defense measures are checked for their effectiveness and revised when necessary. The peasantry presents itself as a united force, strengthened and with clearly defined interests. Members of the upper class begin to withdraw restrictions and controls. An effort is made to form a partnership without destroying the hierarchical order and both sides are pleased with the results. The parish captains responsible for selecting and outfitting new infantrymen are placed under the authority of aristocratic commanders. In order to avoid the favoring of local interests, all parties are required to help those in other quarters should the situation be threatening. Rules of behavior are clearly defined in order to keep the newly gained power on the right track. On the one hand, suspicious foreigners and possible spies (Turks and Hungarians worked with scouts who ferreted out suitable targets for attacks) should be stopped immediately. Otherwise, the use of force is strictly prohibited. The power monopoly of the nobility is not to be questioned - only in the exceptional case of self-defense.

This new-found identity has consequences: in 1470 when Emperor Frederick levies a special tax for military purposes the peasants refuse to pay. Their own preventive measures to protect the country, all the construction and guard duty is enough. They have done their share.

Remarks about the “Plagues of God” in Styria are based on:
Dorothea Wiesenberger, Türken, Pestilenz und Heuschrecken;
also: Die Steiermark - Brücke und Bollwerk;
catalogue of the Styrian Provincial Exhibition 1986, Graz 1986
(henceforth as B and B 86)
B and B 86 also quotes the chronicle of the monastery in Göß.
About the first attacks of the Turks und Hungarians and ensuing reactions see:
Harald Heppner, Das Vordringen der Osmanen in Europa;
Peter Jaeckel, Organisation des Früh-Osmanischen Heeres;
Alois Ruhri, Landesverteidigungsreformen im 15. Jahrhundert, and Neue Wege der Heeresaufbringung;
all B and B 86.

The accounts about the Hungarian Wars and the Baumkircher feud (the incidents in Fürstenfeld as well as the “Minoritenpredigt”) are based on:
Roland Schäffer, Die Baumkircherfehde;
Roland Schäffer, Kaiser Friedrich III. und die Ungarn;
András Kubbinyi, Ungarn und die Türkenabwehr bis 1526;
all B and B 86.
I have taken the quotes about Emperor Frederick III from:
Alphons Lhotsky, Kaiser Friedrich III.,
also from: Aufsätze und Vorträge 2, Wien 1971;
Philippe Braunstein, Annäherungen an die Intimität;
Philippe Ariès and Georges Duby (editors), Geschichte des privaten Lebens 2,
Frankfurt 1990.

Inner Realms - Outer Realms

April 27, 1998
Went to the Zeughaus
to read the papers.

"The crisis in the Serb province of Kosovo
has led to a dangerous escalation
in the relations between Albania and Serbia.
Also critical is the situation in neighboring Bosnia
where for the past days Serb and Croation nationals have been
attacking members of the opposition folk groups
who are attempting to return to their homes.
Yesterday, under the protection of SFOR - troops,
200 Serbs were evacuated from the Bosnian city of Drvar
to the Serb Republic, after Croatian troops had set fire to their houses the day before."

Centuries of war and terror are impossible to erase from human memory. It is easy to analyze how a history of violence can lead to violence in the present. What it really means, however, when generation upon generation live in the heart of a war zone, and perpetuate this war, is only very difficult to conceive. Former Yugoslavia represents such a zone of conflict where two worlds clash brutally . The forefathers of most of today's inhabitants came here fully conscious of the war, for military purposes. This must have consequences. And the sight of the Zeughaus arms, all the spears, lances, swords and pistols, becomes less idyllic, less reassuring, after reading the daily reports from this civil war zone. All the slayings and arson correspond with terrible accuracy to the descriptions of vio-

lence carried out in the same places four hundred, or three hundred, or two hundred years ago. Accordingly, one's distaste but also one's respect for the historical value of the Zeughaus grows with each moment of present-day reality.

New Authority

The development of the war against the Turks and the history of the "Holy Roman Empire of the German Nation" may be defined in terms of political domains, a division of inner and outer realms.

When Frederick II died in 1493 at the then biblical age of 78, his son Maximilian was just leading an army of mercenaries against the Turks. The "last knight", as he liked to be considered, was equally often described as "father of the mercenaries". These paid professionals, elite troops of unscrupulous, war-experienced men are largely responsible for victory or defeat. Although Maximilian realized this, in his books like the autobigraphical *Weißkunig* he continued to propagate the old dynastic war order, as if battles were still being fought between noblemen and not between peoples.

At the turn of the century around 1500, various rivalries and political-philosophical counter-movements were competing for power. The vacuum of authority promotes greediness with various forces attempting to make advances. For a moment the situation seems open. On the one hand are the estates and wealthy cities which notice that the Emperor's central authority is clearly weakened. The economic superiority of the cities grows, their self-assurance as well. And thus in 1488 the citizens of the city of Bruges hold Maximilian captive for three months, forcing him to make far-reaching concessions regarding their sovereignty. Maximilian does not honor his promises, however the audacity of the situation speaks for itself.

On the other hand is the newly inflamed dynastic thinking of the ruler. Even Frederick had considered himself, due to his imperial status, as an excellent monarch - he simply lacked the financial means to fully exercise his authority.
Maximilian also continues to regard himself as "chosen". His arrogance is legendary - and it has genealogical roots. For in fact, the biggest successes of this man, proclaimed in 1508 in Trient as "Appointed Roman Emperor" - the Imperial Coronation did not take place in Rome as Venetian dukes denied Maximilian the right to pass through their territory - are a result of family politics. Through his marriage to Maria of Burgundy, the House of Habsburg acquires the Netherlands. In addition, Maximilian manages to secure Frederick's earlier claim to St. Stephan's Crown and Bohemia. The marriage of his son, Philip to Johanna, Infanta of Aragon and Castile, should finally enable the annexation of Spain. Marriage as the foundation of an empire "on which the sun never sets".

The importance of genealogical relationships is not only evident in the strategically planned marriages - whose advantages nonetheless most often had to be strengthened and confirmed by force against competing rival claims. Genealogical or family trees which trace the family's past are also set up to provide proof of legitimacy for future generations. Even the early medieval elevation of the Habsburgs - who did not originally posses noble privileges - to royal "archdukes" was based on a historical falsification, the "Privilegium majus" of 1358. This declaration, which allegedly dated back to Julius Caesar and other Roman rulers and which intended to grant Austria special privileges, turned out to be the inept forgery of a "jackass", according to the humanist Petrarch whose evaluation was based on a study of the language. This, however, could not undermine the legitimacy of the document. The privilege remains valid.

Emperor Maximilian also had family trees drawn up to trace his ancestral origins back past antique Rome to the legendary city of Troy. He is by no means alone in this practice. Many large families of the day had their history traced back to the Trojans in order to lend their economic and military authority an additional genealogical dimension. And - most interesting of all - the Turks make no exception. After his conquest of Con-

Hussaric Cuirass 1580–1590

stantinople, the Sultan writes to the Dukes of Venice: "Why should we wage war when we are indeed brothers? For the Turks, as all know, have emerged from the burning city of Troy. They are descendants of Turcus who is a son of Priamus, just like Aeneas. "Not even a Turkish sultan wants to be excluded from the calculation of ancestral lineage. A ruler, whose military strength and religious conviction would seem to provide him with enough self-confidence, nonetheless competes with his foe in an intellectual battle over the nobility of origins.

Ruler and Folk

Ideas and actions derived from considerations of genealogical privileges must necessarily conflict with the claims of estates and cities whose self-confidence is built on economic reality. The genealogical argumentation is based on ancient relationships, as if no break between the mythological past - which was purely literary - Roman history and the monarchal presence exists. Thus the justification of sovereignty is founded on literary and philosophical concepts which are so mighty that even the Turks want to make use of them.

The gap which opens up in this period is enormous. For at the same time that the Emperor lets his lineage be traced back to Aeneas, he must borrow exorbitant sums from the citizen Fugger in order to finance his claims. At the same time as these mythically elevated descendants of ancient heroes secure entire empires through marriage and propagation, their subjects face the struggle of survival with a new self-confidence and the decision to "help themselves". They no longer feel protected by the "fatherly monarch" whose genealogical justification includes the ancient patriarchate and its tradition of protecting the dependants in fairness, austerity and mercy.

Society is hurled apart; various groups are possible winners. Those who are most underprivileged also try to advance. Peasants and laborers recognize the fact that others make claim to the wealth they are earning. The conflict is resolved by violence. Draconic measures are taken to insure the status quo. Peasant revolts are crushed by bloody massacres, assertive mayors of up-coming cities land on the scaffold. In the next 100

Breastplate
of Field Armor

Early 17th century

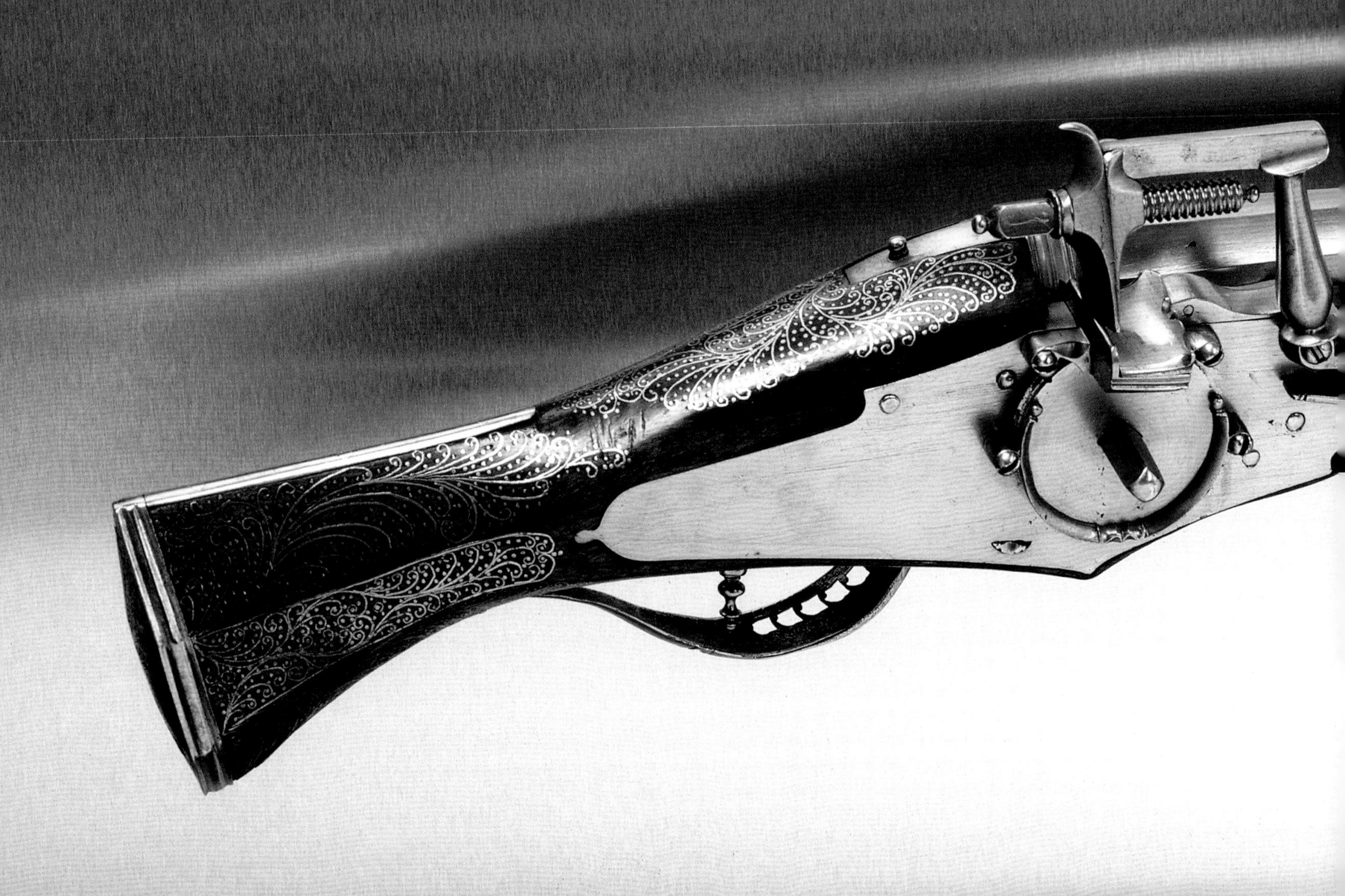

Wheel-lock Pistol	l	61,1 cm
Ferlach, first half of	w	1,54 kg
the 17th century	cal	13mm

years power becomes more centralized after every defeat. However this absolutist empowerment is achieved at the cost of any possible independence of the subjects, as well as at the cost of the mythical-religious justification of power.

The sovereignty of the king becomes more and more dependent on force - which also explains why the state then eventually "devours" its monarchs. The absolute monarch constitues a new form of rulership - his justification is one of repression.

Whereas this repression spreads internally, the war from outside draws closer. It is now necessary to defend oneself against someone foreign, someone different: it is the beginning of the war of races and peoples. Michel Foucault describes it as follows: "In this story of races and their continual confrontation both within the law and outside the

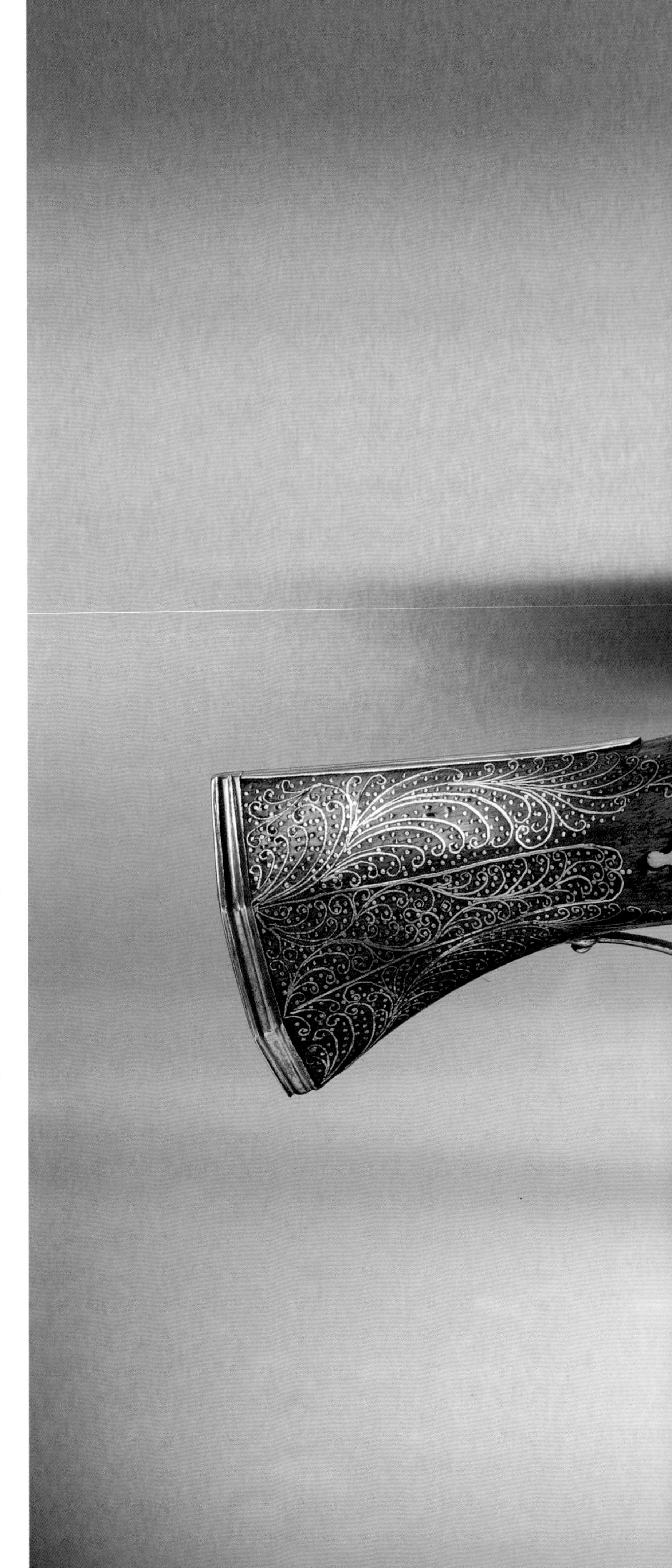

Wheel-lock Pistol	l	61,1 cm
Ferlach, first half of	w	1,54 kg
the 17th century	cal	13mm

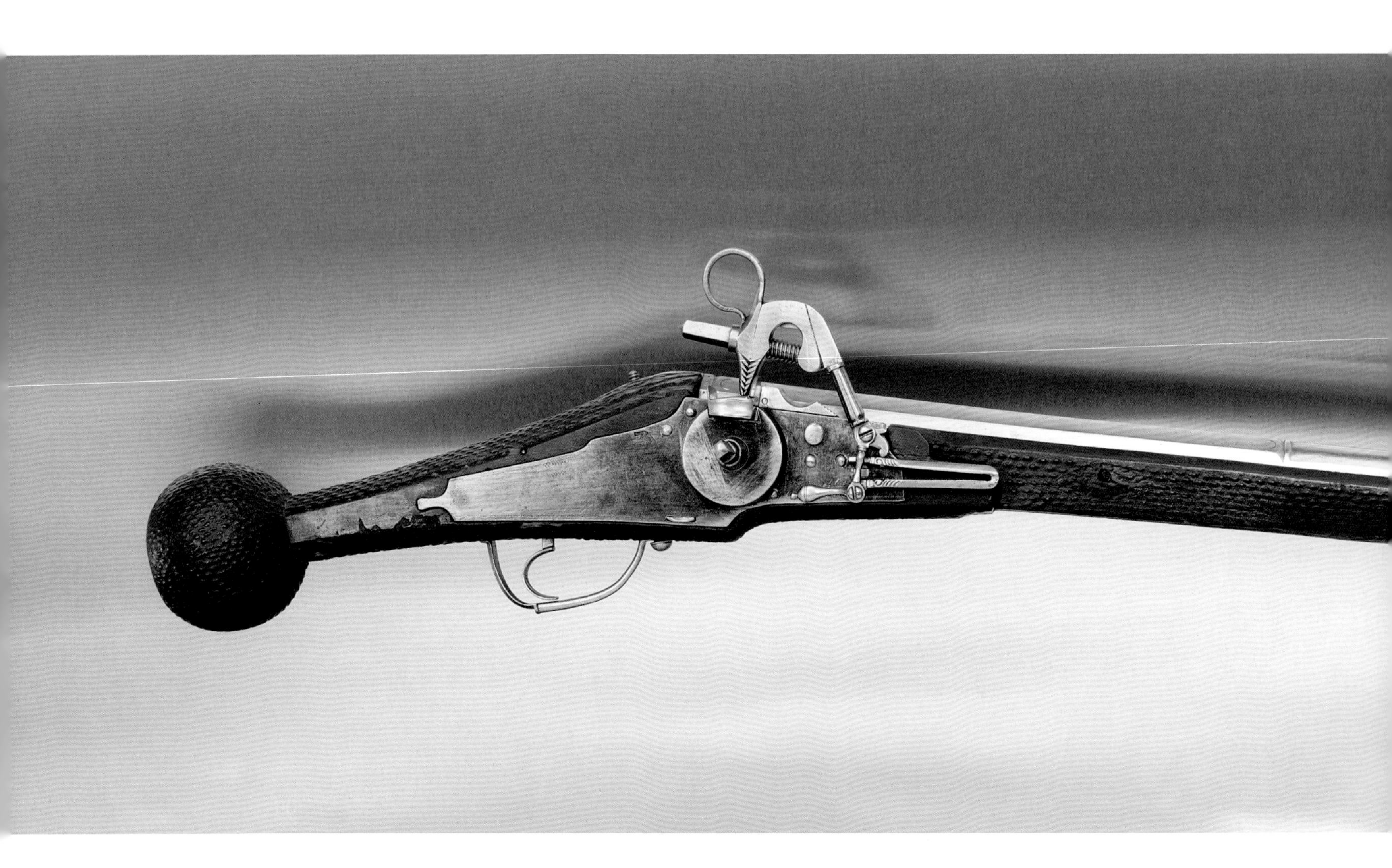

law, the once prevailing dialogue of sovereignty, the identity built up between the people and the monarch, between the nation and the sovereign, disappears. In this new type of historical discussion and practice, sovereignty no longer unifies everything to a single nation or state. Sovereignty now has a special function: it does not unify, it subjugates. " This history of subjugation ranges from Maximilian's ruthless treatment of mayors and local rulers who dared to assert their independence, to the final banishment of Protestant nobility from Austria's inner territories (the function of the Reformation with regard to "inner" and "outer" will be discussed later) which ended in the Thirty Years' War.

Wheel-lock Pistol	l	74 cm
Southern Germany	w	1,73 kg
Late 16th century	cal	13 mm

Common Enemy

War approaches the outer boundaries. Now that the great feuds of the late Middle Ages are over, the new threat comes from outside. Though it was once possible to overpower medieval troops of knights by the sheer number of simply armed soldiers, the tides of war technology have now turned back. Even the Swiss owe their success, in addition to the unconditional efforts of all soldiers, to well-trained war specialists. In the long run it becomes impossible "to help oneself" against sharpshooters and fast-riding cavalry. Military defense carried out by peasants under command of the

Wheel-lock Pistol	l	74 cm
Southern Germany	w	1,73 kg
Late 16th century	cal	13 mm

estates is no longer adequate, the self-confidence of these simple soldiers no match for the new military threat. For not only the Emperor but the Ottoman aggressors as well benefit from the development of firearms. Maximilian bases his success on the use of mercenaries and the effectiveness of newly developed canons. Albrecht Altdorfer's illustration of the *Ehrenpforte* (Triumphal Arch) - a pictoral triumphal procession for which in reality the financial means were lacking - portrays the Emperor surrounded by new war equipment, with the following caption: "He has devised the most horrible of weapons and made them possible at high expense."

In the course of the new centralization of power, the Austrian defense system is reformed as established in the "Innsbrucker Defensionslibell" of 1518. The estates of Upper and Lower Austria, Styria, Carinthia and Carniola (Slovenia) sign a defense regulation which, in addition to local measures, take a further decisive step toward centralization and political unity. Provision of mutual support in case of war is agreed upon.

War councils headed by a commander-in-chief are established and fieldmarshals, captains and lieutenants are appointed to organize the defense

Stock of a Hunting		1600–1625
Rifle	l	138,8 cm
Southern Germany	w	1,18 kg

in the provinces. And finally it is stipulated that in case of defense action the Imperial Chief Field Marshall, who is personally appointed and paid by the Emperor, has the final authority of command. This builds the foundation for centralized military leadership, which will be strengthened and modified in the future. After the division of the hereditary provinces in 1564 (the "empire on which the sun never set" was no longer to be governed as a whole) the estates were reigned by sovereign princes whose influence on the conduct of war continued to increase, although this new division seemed for the moment to contradict the idea of centralism.

The Disaster of Mohács

"I doubt that any Turk is so cruel, that he can wish the Christians more harm than they inflict upon each other." This opinion of the humanist Erasmus from Rotterdam bears witness to the time in which Emperor Maximilian's grandson, Charles V, occupied and plundered Rome with an army of hired troops, and even considered declaring the Pope a heretic and executing him. The *Sacco di Roma* (sacking of Rome) which ended in 1527 with the reinstatement of a humiliated Clement VII, who was then

Stock of a Hunting Rifle		1600–1625
Southern Germany	l	138,8 cm
	w	1,18 kg

Wheel-lock Pistol	l	50,3 cm
Nuremberg 1577/78	w	1,67 kg
	cal	13 mm

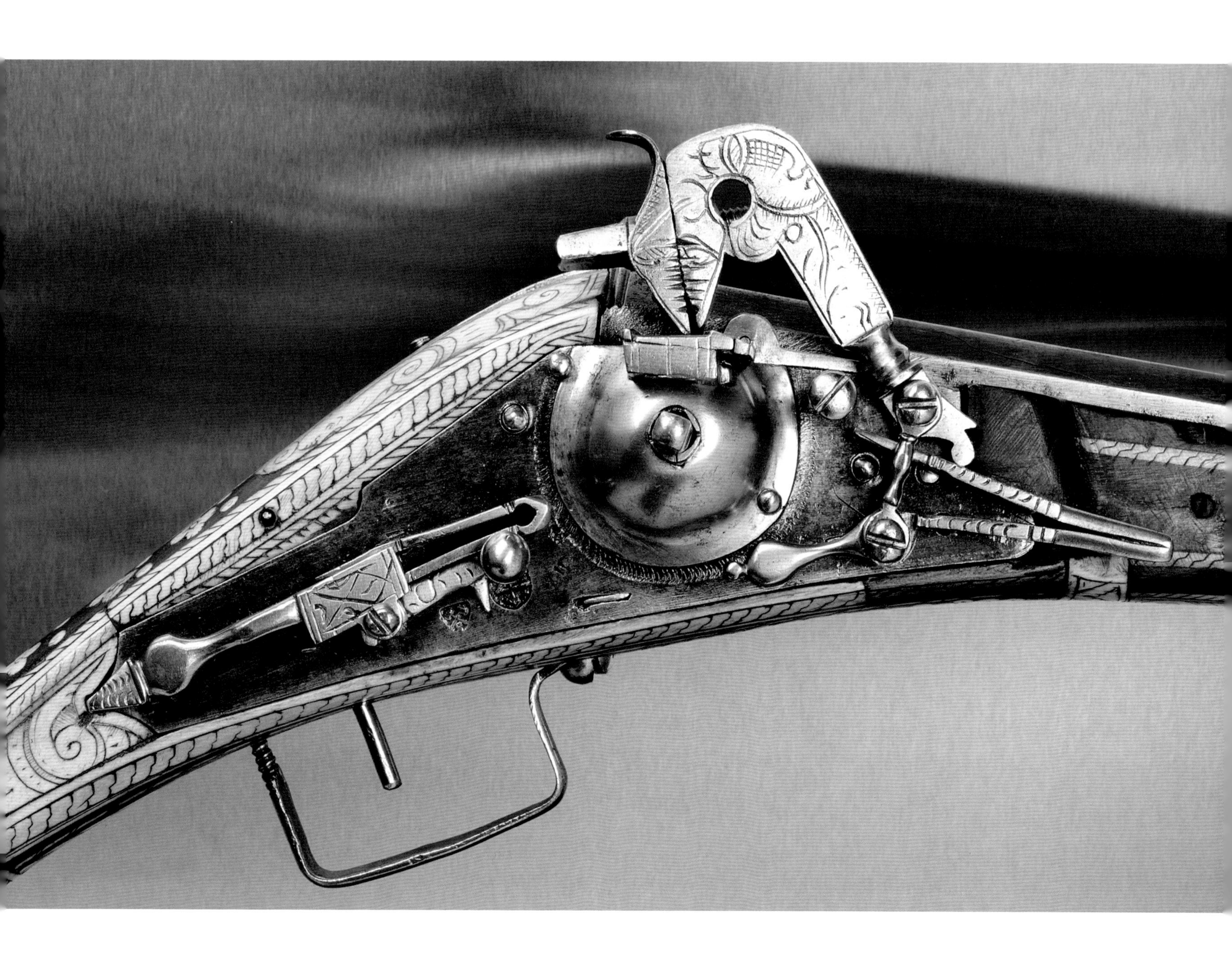

Wheel-lock Pistol	l	61,1 cm
Ferlach, first half of	w	1,54 kg
the 17th century	cal	13 mm

Wheel-lock Pistol	l	60,6 cm
Ferlach, first half of	w	1,50 kg
the 17th century	cal	13 mm

obliged to crown the new Emperor in Bologna (Charles is the appointed Emperor since 1519) can be seen as symbol of the collective atrocities committed during Charles V's reign.

At the same time as an "upstart monk" by the name of Martin Luther declares before the Emperor and the Imperial Diet of Worms that he is only bound to his conscience and the word of God in religious questions and not to the curia (papal court) in Rome, at the same time that Luther is condemned by the Emperor but protected by Frederick, Elector of Saxony, King Francis I of France seals a pact with the Turks to put pressure on the Habsburgs. At the same time that Hernando Cortez and Francisco Pizarro with a handful of well-equipped soldiers and countless Indian helpers annihilate the Aztec and Inca Empires in an unprecedented case of genocide under the pretense of "converting the heathens", the former monk Luther is translating the New Testament while held captive at the Castle of Wartburg in order to free God's word from the monopoly of the Roman church, and make it available to as many people as possible. And at the same time as citizen and peasant revolutionaries, inspired by the theories of the Reformation, try to claim their own freedom by force, exactly this reformer Luther curses them as "robbing and murdering bands of peasants" and propagates, in the name of God, their subordination to the primacy of the reigning prince.

In just this period, Sultan Suleiman "the Magnificent" unifies the Ottomans - who have been successful in Egyptian and Arabian territories though worn by internal power struggles - and sets off toward the West on a great expansion campaign. In 1521 Belgrade is taken. And on August 29, 1526, the Turkish army led by Suleiman conquers Hungary at Mohács while the defeated King Louis II drowns helplessly after fleeing the lost battle. As a result, the Ottomans stand directly on the border to

Wheel-lock Pistol	l	61,1 cm
Ferlach, first half of	w	1,54 kg
the 17th century	cal	13 mm

Austria, which in addition inherits Hungary and Bohemia since Louis leaves no male heir. Earlier Frederick III had managed to secure from Matthias Corvinus the Habsburg claim to succession in case of the male heir's death. Maximilian continued to reaffirm this contract, and the marriage of Maria, Archduchess of Austria to King Louis added a genealogical tie - especially since in addition, Archduke Ferdinand I then married Louis' sister, Anna. As such, Ferdinand I was legitimate heir to the throne of Hungary and Bohemia. After defeating the Hungarian rival king, Johann Zapolya, and making numerous concessions to the local estates, Ferdinand was finally elected by these estates in return. Once again the estates of Bohemia and Hungary managed to elicit certain privileges and preserve their independence. Emperor Rudolph II later limits some of these rights especially with regard to religious freedom. And the final defeat at the Battle of the White Mountain and subsequent expulsion from Prague of the Protestant "Winter King", Frederick, Elector of Palatine, at the beginning of the Thirty Years' War ushers in the era of complete Catholic domination by the Habsburgs and their universal claim.

The First Siege of Vienna

Before it comes to the ultimate establishment of absolute sovereignty, however, countless losses are suffered. Alone in Hungary more than 20,000 fell at the Battle of Mohács. And in 1529, Sultan Suleiman marched into Vienna with approximately 100,000 men. The siege lasted about a month but was unsuccessful and the Turks withdrew leaving behind a trail of destruction and pillage. Meanwhile the defeated Hungarian rival king, Zapolya, allied himself with the Turks, together bringing the province of Transylvania under their control. In addition, the Turks installed themselves in East Hungary. Ferdinand I, with the support of his brother, Emperor Charles V, conquered a few territories in West Hungary, thus Hungary was de facto divided into three parts.

In 1532, a major campaign in which the Emperor personally participated, failed to bring the desired results, but instead had devastating consequences for the Austrian peasants. Retreating imperial mercenaries, in

Wheel-lock Pistol Ferlach, first half of the 17th century	l w cal	61,1 cm 1,54 kg 13 mm

want of rich spoils, plundered the Austrian countryside and behaved no less brutally than the Turkish troops which terrorized many Styrian towns while fighting back and forth in the area nearly reaching the city limits of Graz in 1532. Erasmus from Rotterdam meant just such instances - by no means an exception in these years - when he spoke of cruelty beyond all religious barriers.

The consequences were disastrous in all respects. Crops were stolen or consumed, houses destroyed and set afire, churches ransacked, people kidnapped and slain - the terror reached inconceivable dimensions. One especially common war practice was the abduction and enslavement of people. The most unscrupulous Ottoman troops were composed of abducted Christian children who had been trained as fanatical soldiers in special military schools. Nonetheless one cannot ignore the fact that Christian opponents employed similar practices, also taking slaves (not only in "far off" America, but in Europe, too) sparing no one, and least of all their own subordinates. This is certainly one explanation for the defamatory propaganda campaign against the Ottomans, aided by the new printing techniques of the day including illustrations and leaflets. The Turks - one more example of placing the blame for war on "external" causes while at the same time strengthening internal control - become the epitome of evil. These "foreigners" are an excuse for still more taxes, for the increasingly gruesome treatment of the population, and finally for a new empowerment of the princes, who are celebrated as defenders against the Turks. Why the Ottomans, with all their military superiority at the time, did not make a more consequent effort at expansion remains unclear. It is certain, however, that they did not intend to wage the kind of religious war insinuated by their Christian opponents. It was not a question of extinguishing Christian culture, but rather of proving the superiority of Islam, not as a replacement, but as the consummation of the monotheistic beliefs of Judaism and Christianity. The peace treaty between Ferdinand I and the Ottomans in 1547, by which the Habsburgs had to accept Turkish claims to Hungary as well as pay reparations to the Sultan, was crucial in symbolic terms: a Christian leader had to recognize the supremacy of an Ottoman.

Wheel-lock Pistol	l	58,8 cm
1600–1625	w	1,39 kg
	cal	10,2 mm

Reaction of the Estates

The indigenous troops recruited by the estates during this period have little chance against the concentrated military force of the Ottomans (or against the well-versed plunderers of the Emperor's hired troops). The Styrian estates react with a variety of measures. At the height of Suleiman's power they forgo the use of the largely ineffective and depleted people's troops and hire mercenaries themselves - financed of course by taxes levied against the subjects. Next, the estates decide on more selective forms of conscription in an attempt to increase effectiveness. The main goal is to train suitable males from the local population as riflemen, thus providing them with professional war skills now vital to military success. For this purpose, every thirtieth man of eligible age is drafted according to the Conscription Ordinance of 1556 issued by the Styrian estates.

The ideal army was considered a combination of well-equipped and trained "elite troops" supported by cavalry, the so-called "Gültreiter", as well as hired professionals and a large number of footsoldiers. The draftees were ordered to appear equipped and armed for inspection - hope was placed in a type of civil armament controlled by the ruling authorities. Furthermore, a new strategy was devised to keep the war as far as possible from one's own territory. A buffer zone was created to intercept and counter enemy attacks on the perimeter - battle was displaced to the outskirts, this time through the efforts of the estates. In order to purge the inner territories of war a military frontier was established, the so-called "Confin" which swept across parts of Hungary, Slovenia and Croatia. Peace and authority at home - unity and inner stability, this was the new course of action which finally proved to be very effective.

In 1555 the Treaty of Augsburg is signed which acknowleges equality of rights to both Protestants and Catholics, whereby the reigning prince of each land may determine the prevailing religion. In 1556 Charles V abdicates and Ferdinand I becomes Emperor. The utopian vision of a world empire which Charles had cultivated disintegrates in the face of political reality. Philip II becomes King of Spain and is not able to deter the emancipation of the Netherlands nor the growing of England. When Emperor Ferdinand I dies in 1564 it does not yet lead to a centralized unity as is the case in both France and England. Instead the ruling power is divided according to a plan which Ferdinand had worked out for his sons and had written in his testament. Maximilian II (who supposedly had a strong in-

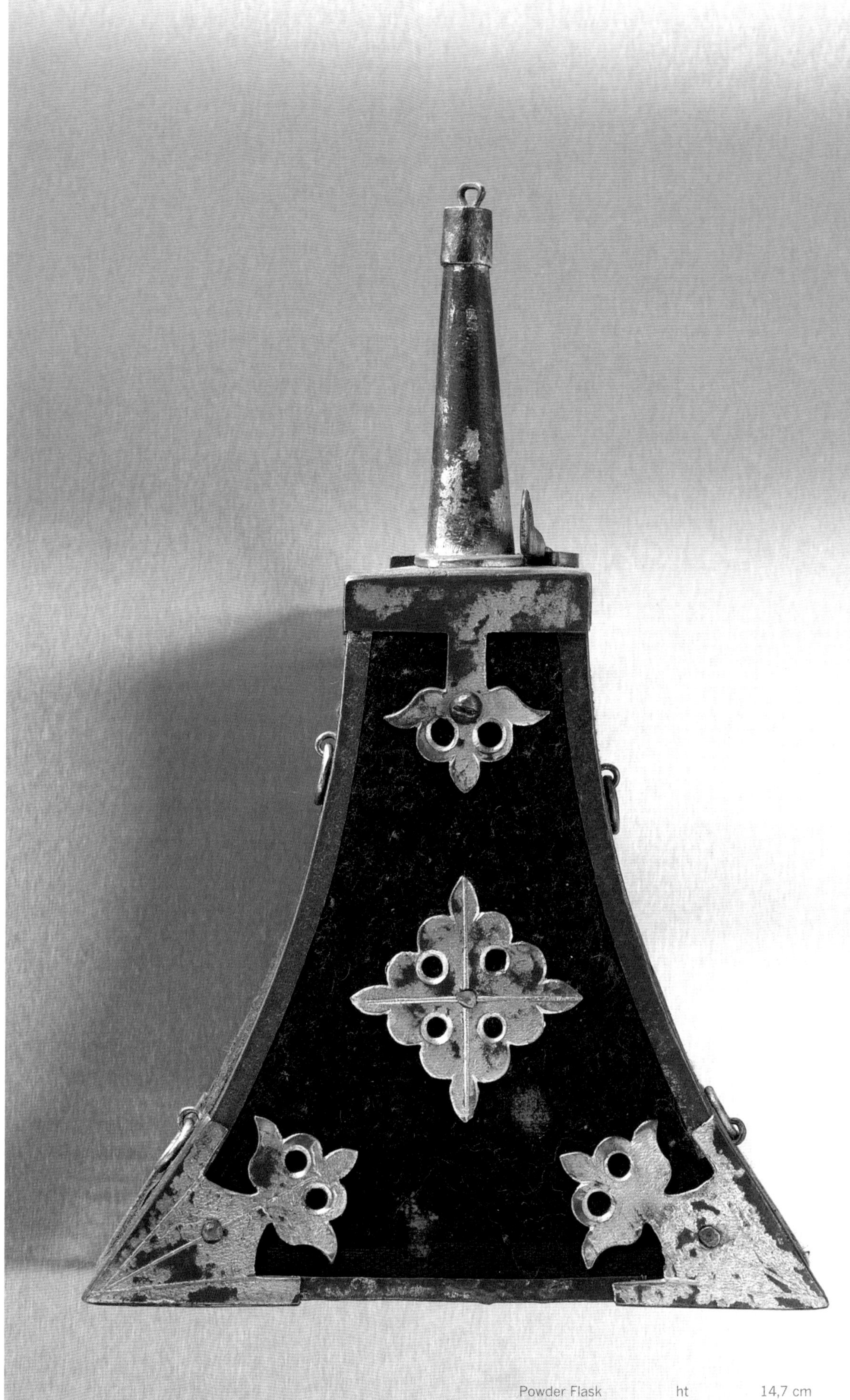

Powder Flask	ht	14,7 cm
c. 1570	w	335 gr

clination toward Protestanism), already King of Bohemia and Hungary in his father's lifetime, is appointed Emperor and receives the region Danubia. Karl II becomes Archduke of Inner Austria (which includes Styria, Carinthia and Carniola) and Ferdinand II becomes Archduke of the Tyrol (with further possessions in southern Germany). For the meanwhile largely Protestant Styrian estates, it is another opportunity to demand additional privileges. In 1564 the estates carry through another revision of their defense system, this time clearly increasing the number of armed civilians in an effort to entirely eliminate the need for mercenaries. Their self-confident position with regard to the archduke is certainly due in part to their profitable administration of the military frontier which has brought enormous economic benefits. In 1572 Archduke Karl II must pacify the estates by granting them the right to religious freedom. The following period shows a great arms buildup in Styria since large stocks of war material are necessary to defend the border effectively. This leads to an arms boom - not only locally - and to the organized planning of military operations and storage of equipment.

The Defense Ordinance of 1575 and the Provincial Diet of Bruck in 1578 both dictate further steps for extensive civil armament, as well as defense and organizational tactics to be employed on the military frontier. Subsequently, the estates of Croatia are stripped of their rights and the administration of the border is formally accorded to the provinces of Inner Austria. Thus the transfer of war to the outer territories is completed, and the rulership can consolidate its inner stronghold.

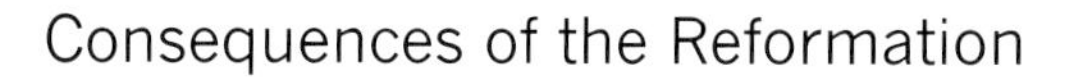

Consequences of the Reformation

Protestantism makes an important contribution to the authority of the "modern" state in the 16th century. Its religious philosophy is one of separation and limitation, thus it newly defines inner and outer confines. Luther objected explicitly to the practice of indulgences which were granted by the Roman church - this protest resulted in his famous "Ninety-Five Theses". The political consequence would mean separation from Rome and thus from ancient Roman history. If the history of the Emperor's sovereignty is linked with Rome and Roman mythology, then it is branded by Luther as a genealogy of decay. Rome is for Luther the dishonest, worldly exterior - presenting the individual a "chimera", or mere illusion of authority. This individual is, above all, personally responsible to God and does not require a "holy" Church as mediator. Together with others he forms a new congregation, a community of faith, which is isolated from the falsehoods of the external world. It manifests no political influence - for exactly this political influence, the mixing of religion with worldly demands, is what was being criticized in the Catholic church - instead it subordinates itself to the political situation. The protected inner realm is purely spiritual, one of faith. All worldly power is relegated to political, not to religious leaders. This rulership must detatch itself from the primacy of Rome and provide for its people. A people defined by its language, thus Luther's translation of the Bible into German. The people having religious, not political sovereignty: this conception of society rooted in Protestantism extends far beyond the medieval order of the estates.

From his religious point of view Luther was right in criticizing the lower classes who were striving for more power. Protestantism, much more than Catholicism, subjects its followers to worldly authority, depriving the individual of all rights of self-determination. Whereas Erasmus of Rotterdam emphasizes the central importance of the individual's "free will" in accordance with the Renaissance tradition of humanism, Luther speaks of the "bound wills" of the faithful who are completely at God's mercy, unable to rectify or save themselves by personal behavior. The obscure activities of the Anabaptists and social activists following the Reformation who attempt to create anti-authoritarian or pre-communist communities (thus anticipating - in a grossly distorted manner - the consequences of future communism) do not fit into the Protestant ideology. Although the Catholic church attempts to blame Protestants for just such misdeeds, Protestantism remains strict with regard to the separation of a religious inner realm and a worldly outer realm - a position which accounts for its strengths as well as its weaknesses.

The first major Turkish Wars are also used as an instrument in the conflict between Catholicism and Protestantism, each group blaming the other that the Turkish sieges and victories are signs of God's wrath. Thus Suleiman's attacks are for the one side God's punishment for the degeneracy and spiritual remoteness of the Catholic church, and for the others the re-

Helmet with Visor 1600–1625

sult of the sacrilegious assault against the true salvation of the old faith. In any case, the political implementation of the religious conflict worked in both directions. And now that it was a question of carrying out the Counter-Reformation with increased fervour, it was agreed that the political subordination of the faithful, which is an integral part of the Protestant doctrine, would also be useful to the Catholics.

Everything foreign - whether the Turks or those in Rome - is defined as one's enemy, as a threat to one's essential being. The compromise formulated in the Treaty of Augsburg, "Cuius regio, eius religio", reaffirms once again the relationship between *inside* and *outside.* Although the individual provinces are small (compared to England or France) and not considered as national entities, the local religion is nonetheless politically important and has a quasi-nationalizing function. Centralization does not follow the same course as in France, but within the individual domains of power, it functions according to the same principles. And it leads to absolutism; for the Protestants do not possess a body of leadership rivaling for theological command, and the Catholics have, during the "Sacco di Roma", overthrown theirs and reduced it to a mainly religious function.

Border Zones

Where is the border finally drawn? Where does one's "own" end and the "other" begin?

To quote Foucault: "The Middle Ages certainly did not know that they were the Middle Ages. But what I mean is: they did not know that they were not antiquity, that they were *no longer* antiquity." This is especially applicable to the border area of Croatia. If one reads the justifications of Serbian and Croatian nationalists in the war following Yugoslavia's collapse, one could even propose the theory that the borders established in

Breastplate of Field Armor   Augsburg   c. 1570

antiquity are still valid. It is a fact that today's conflict, with its antique-mythological connotations, has never been resolved. Within the bounds of former Yugoslavia there has never been a chance to relocate war to outside territory. For generations, this has been the heartland of war, taking the full brunt of clashing interests; a war trench which was conceived, planned and occupied as such. The "Windisch (Slavonian) Military Frontier" is the realization of the displacement of war.

The genealogical conflict which can be traced back to antiquity is obvious. After the division of the Roman Empire into a western and an eastern realm, opposing theological interests quickly developed. The victory of Roman Catholicism in the West and the triumph of the Greek Orthodox in Byzantium caused a religious split which escalated into hostility and an embittered struggle for legitimation. The schism of 1054 brought on the final separation of the two Christian churches. The division line was drawn through the Balkans, whereby the Ottomans with their Islam soon appeared as a third competitor for territorial and spiritual predominance. Rome sends missionaries to Croatia; in 924 Prince Tomislav is crowned King of Croatia in Rome - thus Croatia becomes the first Slav empire of Roman Catholic faith. This royal empire is actually "frontland", bordered on the north by Carniola (Slovenia) and to the south by Venetian territories as well as the Kingdom of Serbia. And Serbia is, since the late 13th century, completely in the hands of Byzantine Orthodoxy. This line of demarcation between West and East is not an impermeable border but a contested zone. Crusades, as well as Venetian, Byzantine and finally Turkish wars contribute to a history of unbroken violence, plundering and destruction. Traumatic to this day is the defeat of the orthodox Serbs by the Turks in the 1389 Battle of Amselfeld. Seventy years later, after the conquest of Constantinople, Serbia and Bosnia finally become Ottoman provinces. And Croatia becomes part of the Hungarian Kingdom.

The Ottomans' behavior plays a decisive role in the internal development of the region. Although forced conversions do not take place in the area, many people adjust rapidly to the new rulers and embrace Islam. In Istanbul, formerly Constantinople, which becomes the new capital of the Ottoman Empire, the Sultan also places the Greek Patriarchate under his protection. As a result, the Greek Orthodox religion manages not only to hold its own in the border areas, but also improves its position with regard to other local orthodox churches - before these regional groups are finally engulfed by the Greek Orthodox Church.

It is in this region - burdened by both past and present - that the Habsburg princes and the estates place their front line of defense. For this purpose, the Croatians must be stripped of their sovereignty. When Ferdinand I becomes King of Bohemia and Hungary, he thus becomes king of Croatia as well. He then proceeds to rigorously deprive the Croatian estates of their rights and erect a permanent frontier consisting of a ring of fortresses headed by a border commander in Varaždin, who in turn is subject to the rulership in Graz. In order to hold and secure this frontier, the line is then "straightened" or "corrected", which means that certain fortresses and regions in the Croatian heartland are simply abandoned to the Turks. Military fortifications are demolished and the defenseless Christian population is deserted. The result is a "no-man's-land" of relentless terror, whose volitale character has consequences to this day.

For information about Emperor Maximilian:
Brigitte Vacha (editor), Die Habsburger, Graz 1992.
The letter of the Sultan from Istanbul and the theories of Michel Foucault are quoted from:
Michel Foucault, Vom Licht des Krieges zur Geburt der Geschichte; Berlin 1986.
About armament by the estates, civil armament, the Battle of Mohács and the frontier line see:
Alois Ruhri, Neue Wege der Heeresaufbringung in der Steiermark;
Franz Otto Roth, "Initiativen" Ferdinand I. nach der Katastrophe von Mohács;
Franz Otto Roth, Die Habsburgische Länderteilung;
Franz Otto Roth, Vorfeld im Osten;
all in B and B 86.
For the history of the Yugoslavian conflict see:
Michael Weithmann, Krisenherd Balkan, München 1992.

Gallery of Faces
The Fear and Fascination of Facelessness

Knightly Troops

Wartime

The knights of the Middle Ages were men of noble birth. Though mass medieval troops presented, and intended to present, a fearsome sight, they remained groups of individuals who fought man to man to defend their own - and God's - honor, displaying boldness and courage. The confrontation was close, direct - the individual protected by kilos of iron. A long mail cape with interlocking metal rings was worn into the 12th century. Later metal plates gave additional protection, and finally massive armor was employed. The field of vision was limited to slits in a closed helmet which came into general use in Europe after 1200 - the knight totally isolated from the outer surroundings so that a page was needed to warn of side or rear attacks. Heavily armed and protected, combat was a physical, bodily effort for men of nobility. War was a matter of the ruling class - common folk and farmers being forbidden, sometimes under penalty of death, to carry weapons.

Tales of knightly battle have been handed down primarily in literary form. The epic poems of the Nibelungen and King Arthur with his Knights of the Round Table describe battles and heroes, acts of valiance and chivalry, tell of what is honorable and brave, what the nobelmen did and what was scorned upon. Usually a duel began on horseback, each armed with a lance as they rode against each other. However, the main combat took

place on foot, with sword, mace and shield, an exhausting exchange of blows till finally one collapsed.
Man to man, direct and relatively slow - the speed of the attack determined by the performance and power of the horses. Armor provided protection from blows, cuts and thrusts; the opponent was wounded or killed - not at a distance, but face to face, an equal. The confrontation - in accordance with the epic code of behavior - was a ritual.

Every society creates laws to govern the use of force: both to legitimize action and as a moral code for controlling power. For the unleashed fury of war brings not only the desired outward effects, but inner dangers as well: the danger of internal instability and the possible spreading of chaos to all warring parties. Gruesome descriptions of deeds carried out by revolting robber knights against the ruling order during the period of the Interregnum may validly be compared to the looting and pillaging of soldiers at the end of the Thirty Years' War. All of these men at war abandoned even a minimum of social order and set up their own laws. With one important difference: the medieval robber knights were well-known individuals; the murdering bandits of the 17th century were a bunch of nameless outlaws. Even war reflected the basic change of attitude which manifested itself between the Middle Ages and the early Baroque period. And yet, the degeneration resulting from war also prompts repeated efforts at consolidation. Violation of martial law remains a strict taboo, a fact well-illustrated by the epic literature of the Middle Ages.

The question can be legitimately asked as to whether not all means in war are justified to achieve the desired ends. Indeed, the existential threat of war almost demands a break in convention. (Chroniclers described with horror how King Peter of Aragon, in the war against the French at the end of the 13th century, ordered the poisoning of watering places thus killing 30,000 horses.) However, this would be underestimating the strong moral ideology of the last centuries. In 1264 when Archduke Frederick of Austria ordered archers to march against the Bohemian troops of Wenceslas I, his opponents vigorously protested: “Oh, Lords of Austria, surely you are noble heroes. You should wage an honorable battle with us, a gallant match with swords and for the sake of maidens' favor. Instead you fire bolts which penetrate our horses' iron coating, felling us to the ground. This is, indeed, dishonorable combat. He who gave you swords shall be cursed; and he who blessed your shields shall never find salvation. Rather should he have blessed a quiver of arrows - this pagan art would serve you well.”

Long-range Weapons

Distance as “mortal sin”. The long-range weapons, longbow and crossbow, are considered unchivalrous, dishonorable. The longbow was first employed successfully by the Normans at the Battle of Hastings. Shortly afterwards Christian soldiers experienced the devastating power of longbows and crossbows on their Crusades. In 1139, the 2nd Lateran Council of Rome outlawed the use of these weapons against Christians: “We forbid furthermore, under threat of banishment, that the deadly and ungodly art of crossbow and longbow shooting be used against Christians or those of faith.” But exactly the crossbow which enabled long-distance accuracy, became a lethal weapon during the High Middle Ages.

A heavily armored medieval knight was also relatively defenceless when faced by armed commoners. Pulled off his horse and unchivalrously beaten - a bold group of simply outfitted footmen were certainly capable of this once they had enough self-confidence or fear to overcome any mental reservations, realizing that the nobleman was no more than a vulnerable human being. This was proven by the Swiss in their national battle of independence against the Habsburgs. The crushing defeat of the noble knights by the likes of such inferiorly equipped farmers and citizens dealt the final blow to the ideal of knightly warfare. Footsoldiers and a more mobile cavalry were necessary - and commoners as war material. Yet if these simple people suddenly play a decisive role in winning war, then they also gain in value. Promises of freedom and tax alleviations for farmers helping to defend the military border to Croatia may well be viewed in this socio-political light.

A further aspect became apparent during the war against the Hussites in the 15th century. Religiously fanatical Bohemians took advantage of all

Pot Helmet
c. 1500

ht 23,5 cm
w 2,18 kg

Close Helmet	ht	26,2 cm
c. 1500	w	2,10 kg

Close Helmet	ht	26,2 cm
c. 1500	w	2,10 kg

Fluted Armor Nuremberg c. 1510

Fluted Armor Nuremberg c. 1510

possibilities: they built barricades of wagons (defense encampments) making surprise raids possible; they fought unchivalrously since their noble opponents had also broken their word. Under the false pretense of safe conduct and immunity, John Huss was coaxed to the Council of Constance where he was convicted and burned as a heretic.

The question of distance: newly defined by firearms. Canons, rifles and pistols redefined the distance between opponents while improving effectiveness and accuracy. No matter how much Emperor Maximilian, considered "the last knight", lamented the employment of these "dishonorable" weapons, no ban could stop their use. This leads to the development of completely new and more differentiated troops. Large numbers of simple footsoldiers are needed, highly skilled sharpshooters take over specialized duties. Firearms are still far from being suitable for the masses. Great skill is required to use them safely. Paid mercenaries are employed to fill these duties. Nonetheless cavalry riders remain noblemen or specialized marksmen with individual suits of armor.

Uniforms and Uniformity

Individuality behind a mask. Signs - capes, coats, scarves and sashes signalize friend or foe. These are the beginnings of uniforms though the idea of a homogenous army is still distant. Armor is rather considered a second skin, a protective skin shaped to fit the body, imitating movements with joints to facilitate maximum freedom. This explains why the helmet with its slits for vision and ventilation appears as a stylized face. Mimic frozen in iron, aimed to frighten. Many of these faces seem to laugh scornfully. They appear superhuman, monstrous. They hide the feelings of those at battle and feign an unshakeable superiority. Often patterned after imposing figures of the animal kingdom - the beaks of birds of prey or great dog faces - more often reduced to abstraction, like unearthly insects. These faces of iron remain to this day, at the end of the 20th century, both fascinating and fearsome. They still serve as models for the superhuman: from robots and motorcycle rockers to film characters like "Darth Vader" in "Star Wars". Their appearance is intimidating, menacing

Close Helmet	ht	30,8 cm
First half	w	1,82 kg
of the 17th century		

Helmet with Visor 1510/20
ht 29.1 cm
w 2.28 kg

Helmet with Visor 1500–1525

and brutal just as the helmets of the Renaissance heavy calvary were meant to be.
The protective skin of metal dress was also subject to trends in fashion, a metal imitation of the curves of civilian costume. Perfectly fitting protection which functioned as long as the combat was man to man. Only in the Late Baroque period, when armor had lost its protective function and guns were in mass use did uniformity begin to appear. It is hardly an accident that mass industrialization and the age of enlightenment coincide with mass warfare. The Prussian ideal of militarizing society, of an armed people, is responsible for the depersonalization of state and war, setting new standards even before the people's army of the French Revolution. Broken by drill, members of the army have no protective skin. It is also no longer necessary as the individual cannot act on his own initiative. He is reduced by the uniform to a norm - just one in a human formation of interchangeable parts. If a soldier in the front rank falls, he is quickly replaced by the man behind. Individual courage gives way to mass discipline. Indeed, discipline in the strict order of a military system even replaces the defensive mechanisms of the individual. In accordance with the principles of the Prussian soldier-king who claimed that his soldiers were more afraid of their superiors than of the enemy, strict regimentation and drill became the norm and accepted structure of the 18th century.

Michael Foucault rightly points out that modern day training for military service subjects civilians to general governmental authority - a process in which, starting with the period of Absolutism, the authority of individual noblemen was also replaced by the anonymous power of the state. Prior to Absolutism, subjects were integrated into a system of individual responsibility; later the absolute monarch becomes symbol of an invisible, all-encompassing authority in the form of a state power structure. As in the military, the penal code - imprisonment instead of corporal punishment - also develops generally accepted structures, which reduce the individual to one small part in the overall state mechanism. Foucault illustrates his point with examples of jurisdiction. In the days of corporal punishment, a confession of guilt was an important prerequisite for dealing out punish-

Fluted Armor 1515/25
Innsbruck

ment. The fact that confessions were often forced by gruesome torture should not detract from the basic humanity of this position: here individual responsibility is of central importance. Punishment for one's crime is meted out on one's own body. There was no such thing as imprisonment, locking away an individual by decision of the state, regardless of a confession. Only since the Age of Enlightenment is the pronouncement of guilt or innocence relegated to a system, just as it is in the military.

Upon receiving a command, the orderly must completely subordinate his behavior, his feelings, his privacy and his individuality to the system which makes all decisions for him. The conduct of war and act of killing are thus gradually transferred from the realm of personal responsibility to the reflex of obedience. This process, with various modifications, is still the basis of functioning armies today, especially when personal feelings as to the definition of the enemy might interfere with official orders. Members of troops on the border to Eastern Germany recall how their readiness to shoot down border refugees was encouraged by extreme depersonalization, including the relinquishment of all private articles such as letters, photos - even personal underwear - upon entering the unit.

This is the dark side of the philosophy of Enlightenment with its equality of men. Even if the ironclad knights of the Renaissance and early Baroque period which dominate the Zeughaus of Graz appear to represent extreme brutality, they are individuals. Here, one is far removed from modern-day massacre in a "storm of artillery". And yet, they are no longer knights either, no medieval heroes whose deeds are only poetic projections handed down in epic form - although some tournament armor is actually housed in the Landeszeughaus. Tournaments as aristocratic sporting com-

Fluted Armor

Hans Maystetter
Innsbruck 1510/11

bat for the ruling class existed way into the 17th century, though only as ceremonial reminders of the past; a special kind of "historicism". The chivalrous virtue of the Middle Ages, the strict hierarchy of the estates - if it ever really existed beyond the literary utopia of "Tristan" or "King Arthur" - has long since given way to more complex political relationships. One must only be reminded of the many concessions which reigning Austrian princes had to make to the Protestant estates before the Thirty Years' War redefined relations and Absolutism opened a wide chasm between the common people and the ruling class. This turning point of modernity and rise of the masses is marked by the Landeszeughaus Graz.

Metaphorical Physiognomy

The philosopher Aristotle believed the form of human faces to reveal inner qualities of character. If a person's head reminds one of a lion, then the strength and wildness of this animal must also be inherent qualities. Greek philosophers pursued this idea by associating symbolic characteristics to all kinds of animals from eagles to donkeys. This early form of physiognomy, with its double interpretation - from metaphorical animal properties to human characteristics - is one of the earliest attempts to equate external appearance with inner qualities. Over the centuries, these efforts found not only proponents, but equal numbers of vehement critics as well. Whether determined by the ordered properties of planet energy as the Neoplatonic philosophers maintained, or by human physical features as the scholar Johann Lavater tried to verify, the face is always the conveyor of basic information defining the nature of the individual. No matter what standpoint one favors today with regard to physical categorization - especially in view of the experience with National Socialist attempts at racial physiognomy - the symbolic content of the face is undeniable. To have a face means to be recognizable. The face determines the person. And for many ages, the face is a central theme in the history of art.

Helmet of Tilt Armor   Augsburg   1570/80

During periods when the physique was defined in symbolic terms, the face was above all a representation of something. Studies of antique statues of Roman emperors show that despite the apparent individuality of the portraits, an intended posture dominates. The faces of the emperors are uniform, they exhibit functions, express deportment and bearing such as majesty, superiority, or even spirituality and make use of the formal principles of physiognomic gestic. Not that the sculptors did not portray individual personalities; individuality and formalization co-existed harmoniously. The portraits are always both moralistic and symbolic.

The True Face

The private, the "true" face of the individual becomes increasingly important in the arts between the 12th and the 16th centuries. Not that medieval artists were incapable of realistic portraiture - one is reminded of the distinctly pronounced features on the masks of romanesque capitals - they simply lacked the desire to place individual features on the same level of importance as the symbolic meaning. With the "discovery of the self" (Danielle Régnier-Bohler) however, came the interest in illustrating

Tilt Armor Augsburg 1570/80

personal features, at first without replacing formal principles. Philippe Braunstein points out in his essay "Approaching Intimacy" that portraits of kings and rulers of the late Middle Ages and early Renaissance are still mainly "constructions". The concept of a harmoniously ordered cosmos is interwoven with the facial interpretations of Renaissance artists. The Viennese humanist Johannes Spießheimer invented the metaphor of a "four-cornered king" to represent Emperor Maximillian's figure which is built like a church that reflects God's glory. This building-metaphor for the body can be extended to suit other areas. For instance, the eyes of the Emperor were like stained glass church windows emitting a radiant light, according to Spießheimer. "An increasing number of painted and sculpted portraits of royalty in the late Middles Ages allow us to compare and verify the accuracy of the descriptions. Instinctively we tend to trust the artist rather than the chronicler. Meanwhile painting contains a certain ambiguity which certainly derives from social conventions as well as the intentions of the patron," warns Philippe Braunstein.

Even an artist like Albrecht Dürer, whose self-portrait as a nude dating from 1512 reveals an almost drastic clarity, tends to emphasize in his oil

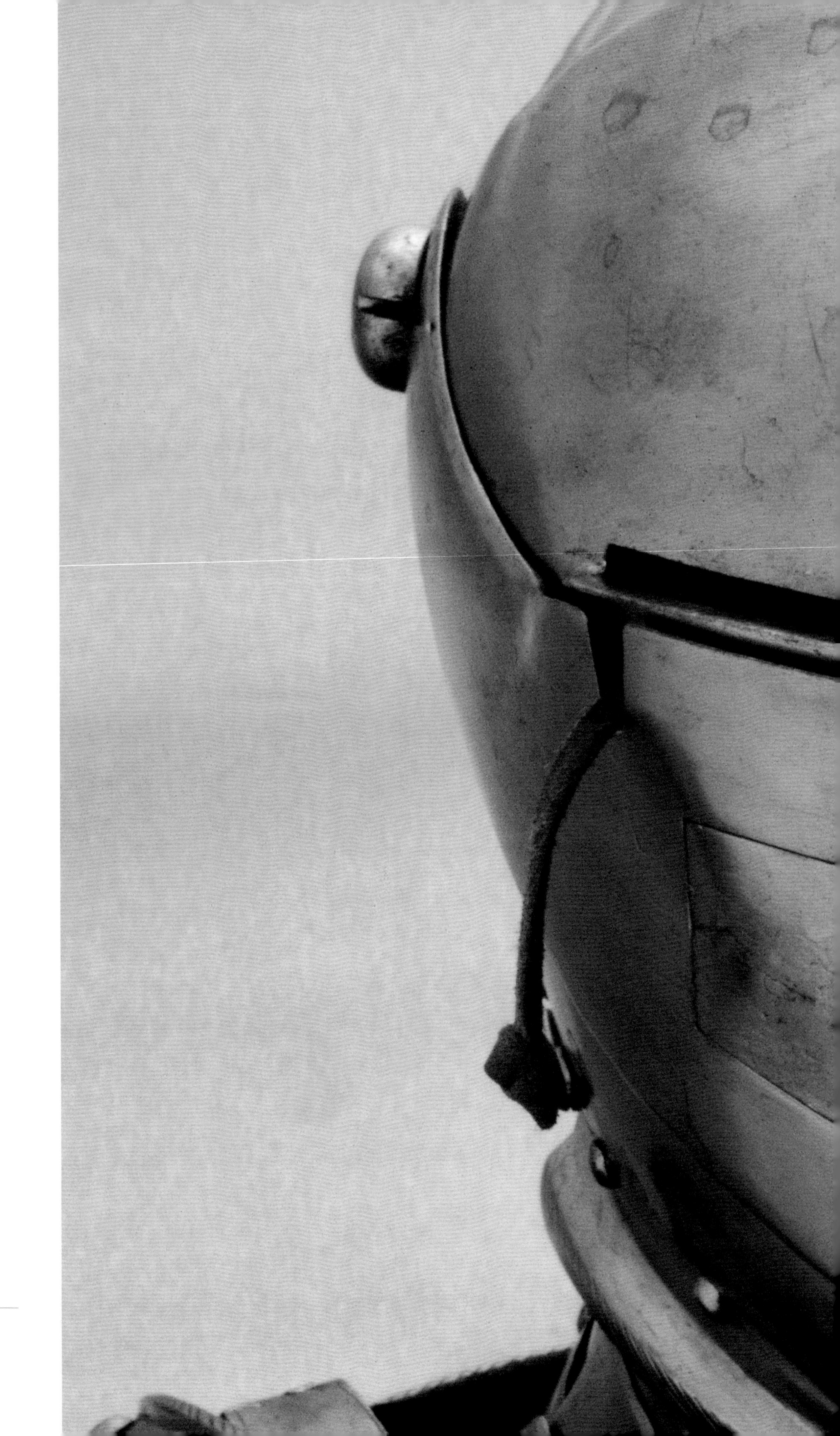

Helmets of Tilt Armor Augsburg 1570/80

Field Armor c. 1600

Field Armor c. 1600

Field Armor c. 1600

Field Armor

Second half of the 16th century

self-portraits a symbolic-harmonic construction rather than sober reality. Whereby this symbolic-harmonic style complies with the contemporary artistic attitude that universal truth can only be found by blending individual features with general rules of harmony. It is important to consider the face as classified somewhere between the principles of impersonal presentation and the beginnings of self-identification when viewing the armor helmets of the Dürer period. The new flexibility of the full-visor helmets which appeared at the turn of the Middle Ages to the Renaissance not only fulfills the necessity of meeting altered battle requirements, but certainly also reflects the new, more discriminating attitude toward the face. The observation that armament is clearly influenced by fashion verifies this connection: if arms are "clothing" for battle which, in addition to functionality, also express symbolic and social connotations of decorum, then the same must be valid for the "facial dress" or the visor. Metallic gestic as a message to the viewer.

Field Armor c. 1600

The Universal Language of Faces

It is a fact that faces speak a universal language. This is achieved by mimical expression and the varied play of facial muscles rather than by symbolic references. Mimicry reveals the state of a person - or how he wants to appear. Experiencing and interpreting facial expressions is one of man's most primitive abilities. This ability is independent of culture (as shown by Irenäus Eibl-Eibesfeldt and other ethologists) and the basic patterns of emotional expression are created in all cultures (even in anthropoid apes) by the same forms. Anxiety and fear, joy and grief find the same universal means of expression in every human face.

It is our fundamental familiarity with the forms of facial expression which, along with touch, is perhaps *the* most important factor in providing security in inter-human relationships. Even very small children respond to one's changing facial expression, showing fear or happiness accordingly. If this level of communication is distorted or completely cut off, people react with visible uncertainty, with fear. Disfiguration of the face presents grave psychological problems not only for the victim, but for his viewer as well. Even when prepared, people confronted with disfigured individuals

Field Armor c. 1635

Field Armor c. 1635

Field Armor c. 1620

report that it is almost impossible not to react to the abnomality with alarm. It is for the most part the foreign and unpredictable quality which actually causes agitation. Habitual "reading patterns" become invalid and a central medium of communication is eliminated.

Having no face means that no change (or only very limited change) is possible. Visual expression is legible by reading movement. Having a face means displaying emotions and fears. Damage to facial muscles or limitations due to paralysis hinder communication and cause alarm. This psychological mechanism is also responsible for the fright incited by masked faces. The close helmets of the Zeughaus serve the function of masks. They show a dehumanized, rigid physiognomy which hides the true face. Emotions are not legible apart from the stylized expression which characterizes the helmet. Whether the soldier is fearful or brave, whether he is sweating and exhausted from the exertion of battle or whether he has managed to conserve his strength - his opponent discerns nothing from surface appearances. The individual has become unapproachable, his personal bearing unperceptible.

Field Armor c. 1600

Field Armor c. 1600

Caricatures and Masks

The function of the mask aims at depersonalizing and standardizing. The animal masks of early cultures served to elevate individuals into the realm of symbols. Masks act as superimposed characters by either exaggerating the ritual - as further development of the religious rite - or by exaggerating in the arts (as in the Greek theater). In Europe masks have been used since the Middle Ages, above all to ward off evil. The grotesque caricatures on Romanesque capitals and facades are dreadfully repelling because they embody dreadfulness. The idea that evil may be frightened off by its own image is founded in ancient conceptions of universal harmony which have been valid for many ages. The masks of the pest doctors of the 15th and 16th centuries with their long bird-like beaks also unite functionality (resistance to the harmful "miasmas" which were thought to carry disease) with symbolism. The frightful appearance of the mask directs, so to speak, the evil outwards; the wearer is not infected, the evil is discharged from the body and blocked off. The wrath of the mask averts the current of negative energy and unloads it - like an electric shock - on the viewer, who recognizes the evil in his own mirrored image, in the reflection of the mask.

The mask is disguise. A second skin which, in a ritual act - theater can still be viewed as such - is pulled over the individual. An eminently symbolic procedure. Of course, the vast majority of close helmets do not possess

Field Armor c. 1600

the distorted character of a "devilish" mask. Stylized, grotesque caricatures on armor are very seldom and even these do not terrify chiefly by means of a shocking appearance. More striking is the sober, almost abstract clarity of the majority of helmets. This does not imply that they had no apotropaic or warning function, but rather that the common "devilish" patterns used to frighten do not apply.

In contrast to cult and theater masks, the helmets are void of symbolism; they do not fall into the category of metaphorical faces which evoke fear. On the contrary, it is rather the complete absence of symbolic identity which makes them frightening. The wearer's faceless non-identity goes beyond the normal realm of theater and archaic rituals, and as such becomes a new and very modern, completely depersonalized outer skin. This accounts for the insect-like, technicized impression these helmets make even today. Which is by no means a contradiction to the new vision of "self" as "individual" which was just becoming popular. Indeed, it was now finally possible to forgo the idea of the "individual" since this had already gained intellectual acceptance. The veiling of one's personal identity in iron completely void of symbolic form is only then truly frightening when the loss of personal identity can be perceived.

The Catastrophe of Depersonalization: Tancredi and Clorinda

In the reflection of the battle helmet, a depersonalized warrior meets his opponent. Transformed by armor into war machines, these individuals nonetheless experience their war character as something separate from their private character. The Zeughaus warrior is still a long way from La Mettries' theory of the man as machine. The truly mechanized man needs no metal skin, for mechanized thought processes are already welded under his own. In contrast, the machine-like warriors of the Zeughaus remain individuals even under the depersonalizing armor. The terrible conflict arising between the metal disguise and the person underneath is a philosophical and literary theme of the day - most accurately described with horror in Torquato Tasso's epic poem, *Jerusalem Delivered.* The Christian, Trancredi, falls in love with the Moslem maiden warrior, Clorinda. One days she leaves her camp dressed in foreign armor. Tancredi encounters her at night but fails to recognize her in the dark, disguised as she is by the helmet. In an outburst of blind hate, they fall upon each other and deliver a fierce exchange of blows. The description of this combat, despite all literary exaggeration, gives an idea of the sheer brutality

required to carry out a battle with swords. ("Darkness and fury remove their habitual skill ... They hit each other with crude, vicious blows of their fists, crash their helmets and shields together", XII, 55,56). Finally Tancredi stabs Clorinda and, after removing her helmet, recognizes his lover. In Tancredi's long lamentation which follows, Tasso describes the personal consequences of deception, using the occasion to elucidate the relationship between truth and illusion, reality and semblance. Whereby the pierced armor of the pagan woman symbolizes her coating of disbelief which is finally opened - (breaking of armor) - by her conversion. For as she lays dying Clorinda begs to be baptized; her friend Tancredi complies. The concealment under armor stands in context with the night when the fateful meeting takes place, whereby the conversion and baptism take place at sunrise. Thus the metaphoric scope of the epic, completed in 1575, includes both nature and human works in a differentiated universal metaphor. The second skin of armor displayed here is a false guise, which conceals the truth and must be overcome. The human and his armor are separable, not one and the same.

Both in full armor, Tancredi and Clorinda see each other as a reflection of brutality. Equal confrontation requires that both parties are equally armed. If one is undisguised, naked in the face of his armed opponent, then superiority becomes obvious and the mechanism of fear takes over immediately. (Another literary source discloses a further example of such

imaginary fear: Cervantes' Don Quixote is haunted by the repeated appearance of the Black Knight - a figure wearing a close helmet, the anonymous and unapproachable archetype of armed evil. Cervantes' stroke of genius is to portray this knight as a mere figure of the imagination, of a man so preoccupied with the crazy ideas found in books of chivalry that he becomes ridiculous. Nonetheless the archetype still functions.) Once two opponents are equally armed, however, they see themselves reflected in the other. Yet this reflection is deceptive and no key to self-knowledge. Dürer's self-portrait, drawn from his mirrored reflection, is just the opposite of this armored view: not obscured but open, not guarded but vulnerable, not rigid but corporeal.

Narcissus in Armor

The "Narcissus pitfall" is applicable to armor.

In his *Metamorphoses* Ovid depicts Narcissus as a cool, beautiful youth who ignores the advances of worshipping maidens. The nymph Echo feels so slighted by him that she degenerates physically, pining away till no more is left of her than a stony echo. Finally a rejected lover curses the youth, wishing him the same suffering he has caused in others. And so Narcissus discovers his reflection in a pool, falls in love with his own countenance which remains ever unreachable, and finally dies of sorrow. That

self-recognition is only possible through another: Ovid sketches this classical philosophical attitude in two interwoven examples, the repetitive Echo and the egocentric Narcissus. The tragic and paradoxical attempt to find a partner in one's own image is responsible for the collapse of oneself.

The "self" disguised by armor finds his counterpart in the disguised "self" of his battle opponent. Here only non-recognition is possible, or rather the recognition of disguise. As such, armed men constitute ideal fighters, the pure essence of war. Each faces a reflected threat, enhancing one's readiness to fight. This form of detachment from oneself is, according to Eibl-Eibesfeldt in his analysis of the violence of war, a basic prerequisite for man's willingness to kill another. Isolation, distance and thwarted communication are, according to the ethologist, a necessary prerequisite of war. They reduce innate and learned inhibitions regarding aggression and minimize one's misgivings about killing another human being. Of central importance is the hinderance of any feeling of sympathy, which in many cases deters the fatal outcome of a conflict. The previously cited literary episode of Tancredi and Clorinda from Tasso is a

classic example. Each time exhaustion or scruple threaten to wear down the combatants, they shout distainful insults at each other, rekindling their hatred. However, the question whether war expresses an exceptional, pathological state of human society is not only negated by Tasso's 16th century literary standpoint. Behavorial psychology also clearly verifies that war serves chiefly to fulfill a purpose. The type and structure of war as an important political instrument will be discussed in depth in the following chapter. However, it is undeniable that specific psychological mechanisms are required to make individuals design and carry out war.

The external iron body is thus a cool, narcissistic body both aggressive and aimed at provoking aggression by erasing individuality. Inhibitions are reduced since one is allowed to attack and kill the external, narcissistic "self". The individual behind the armor is of secondary importance, and as such the iron mask can have fatal consequences.
Not only is the opponent easier to kill, but it is also easier to be killed in return. The armor plating in this vicious circle is not only protection and bodily reinforcement, it also implies self-abandonment and an invitation to destruction.

Masked View

The individual is disguised not only by the metallic mask which conceals his facial features, but also by covering his eyes. Muscle movements, and above all the kind of look, size of pupils and direction of the glance all allow us to evaluate and thus communicate with an individual. This visual opportunity is cut off by the closed visor. The slits, which are kept as narrow as possible to protect the eyes from injury, are primarily an impairment to the wearer's scope of vision. But in addition they prevent eye to eye contact.

Eye movement and width of pupils, contraction of the eyelids and the play of eyebrows are eminently important, though often unconscious signals of human co-existence. Affection and aversion are directly associated with wider, or in the latter case, smaller pupils. Agitated, nervous eye movement signalizes fear, the position of the eyebrows typifies aggression or friendliness. The elimination of this criterion is similar to hiding or freezing facial expression. It irritates, renders foreign and prevents an assessment of the other. It also offers tactical advantages due to the fact that it is impossible to follow the direction of the glance. More important, however, masking the eyes is another form of depersonalization.

This depersonalization contains as a principle strong elements of form, structure and organization. The metallic face of the close helmets presents, in a sense, a picture of composure, emphasizing pure structure by more or less eliminating those emotional elements that can normally be varied at will. This explains why the helmets of richly decorated and etched suits of armor seem to be enveloped by the decor which has "grown" from the breast and arm pieces upwards to cover the entire head. Especially the structural elements, such as the slits for vision and respiration, appear in their abstract and formal alignment to replace the human element with decor. The functional sternness of the iron face, the elevation of the expression to an artifical form serving the desired purpose and effect, signalizes a command of the situation and thus superiority.

This explains why most anthropomorphic traces have been abolished from the visors. As such they surpass the camouflages of ritual masquerades, forming super-imposed structures. Eventual remnants of emotion or gestic which define the animal masks and war painting of so-called "primitive" cultures are reduced or only faintly existent. As a result, these men in armor are elevated to another more important and dominant level of culture - an aspect which is displayed interestingly enough in the wood-

cuttings and etchings used for propaganda purposes during the first phase of serious warfare against the Turks. The Turks are portrayed as savages, men in flattering coats with grotesque beards and mask-like foreign features. Their faces seem to express a rough, uncontrollable brutality, a primitive recklessness. In contrast, the close helmets appear superior, concentrated and above all, powerful. These warriors are not lost in the fury of battle, but stand bravely on the "side of Minerva", deliberate and superior in their actions. In this context the armor plays another role in forming its wearer. The external form not only depersonalizes the soldier, it also gives him confidence and support through the artful craftsmanship and superior design of a culture - a culture worthy of defending and preserving. At the risk of one's personal identity, at the risk of one's life.

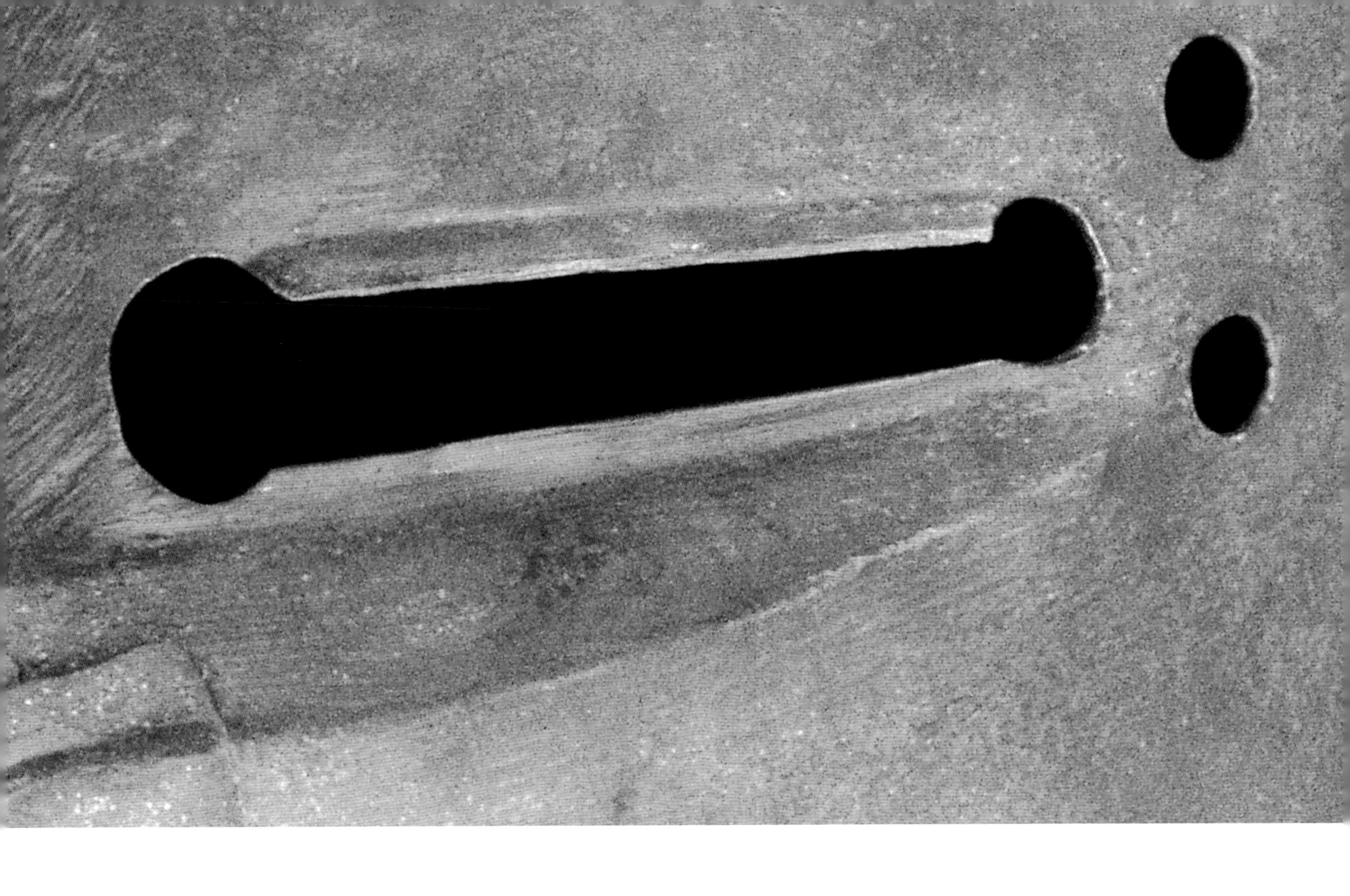

Metallic Figures of Expression

It is instructive to consider which principles of depersonalization are employed when regarding the expressive possibilties of the metallic mask. The most important feature of expression of the 16th century helmets in the Zeughaus of Graz is the visor. The helmets - with their more or less developed combs and narrowed skulls, according to the date of origin - almost never possess anthropomorphic elements. Whether plain, or in the elaborately fluted manner (especially popular shortly after 1500) or with etched ornaments, they always appear abstract. Associations with a rooster's comb may arise in the case of an extremely pronounced comb, however, much stronger than any connections with natural forms is their

pronounced constructive element and the importance of their cooly technical execution. Many of the newer helmets have a peak or brim which provides special protection for the eyes and lends the helmet a cap-like appearance. The riveted visors display a wider range of variation. All must allow for a free range of vision - either by means of slits or wider eye-like openings - and they must make respiration possible. These ventilation slits are usually geometric patterns punched in the metal, arranged in either round or triangular groups. Some are narrow like the eye slits, others are round. Occasionally a wide opening near the mouth allows for breathing, in rare cases there are large vertical slots imitating the nose. These are the basic elements which compose the "metallic face" of a 16th century helmet.

In his standard work, *Man's Face and Mimic Language,* which appeared in 1969, C.H. Hjortsjö schematized the results of facial muscle contractions and analysed them interculturally.

Hjortsjö categorized 24 basic types of facial expression which, although not exactly comparable in every detail, nonetheless reveal that basic structures can be clearly associated with certain emotions. These types of expression are derived from 18 patterns showing various possible combinations of the main human facial muscles. If one compares the schematic structure of the Zeughaus helmet-visors with Hjortsjö's chart it is obvious that none of the "iron faces" conform with an emotional type. Characteristics important to the formation of expression, and which result

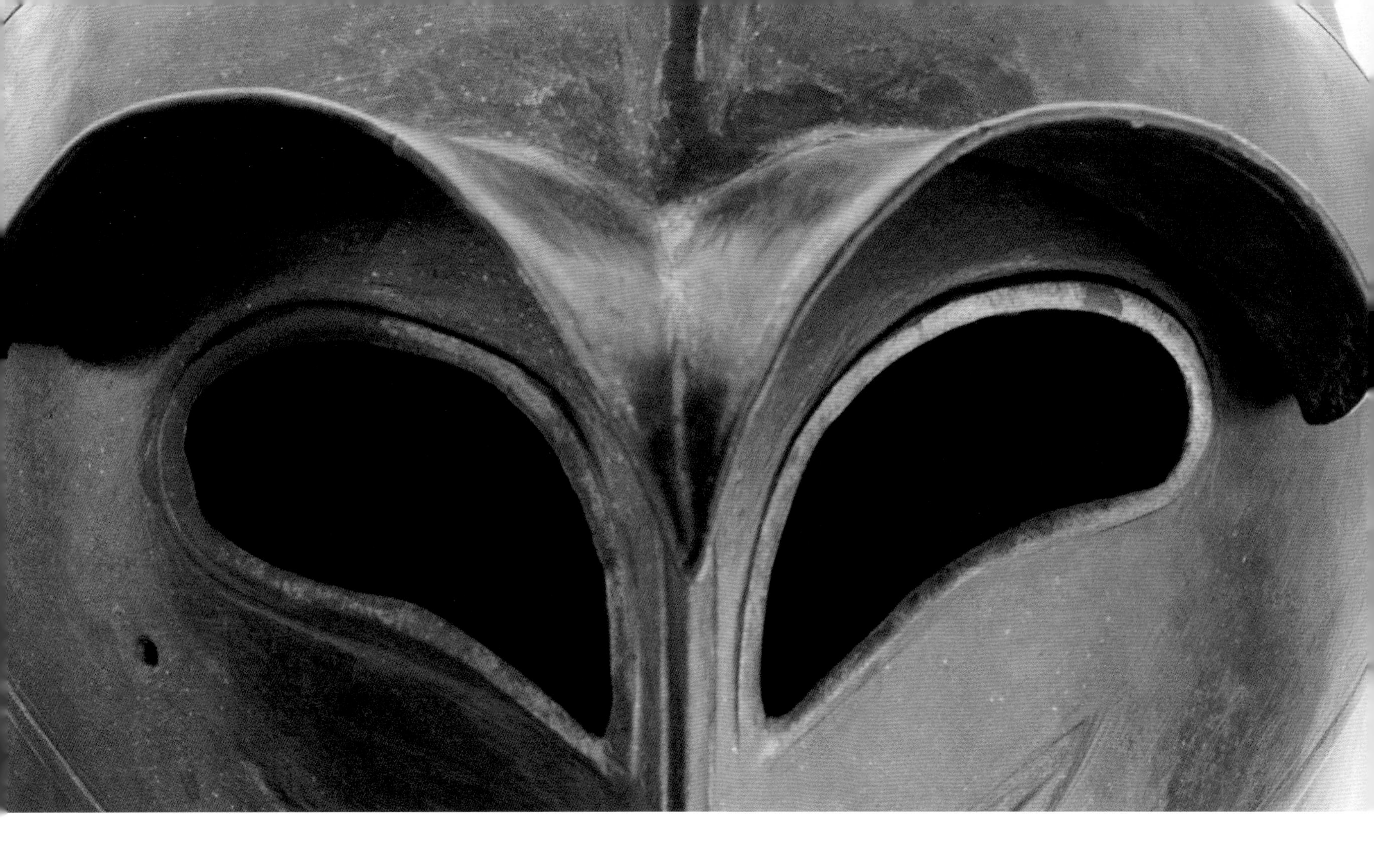

from the movement of various muscle combinations, are absent in the visors. These "masks" do not imitate an emotional facial expression and thus appear without expression, indeed, without emotion. They remain uninvolved, take no part in communication, even in the case of anger.

There are, however, clear similarities to the 18 possible types which Hjortsjö designates as being unable to categorize, those at a pre-emotional stage. The visor faces correspond most exactly to those patterns which display the least emotional coloring. These are the muscle positions evoking indifference, dispassionate calm and distance - borderline areas - which, depending on the angle of the viewer, permit multiple interpretations. The visors mark the frontier between defensiveness and friendliness, between apprehension and relief. They seem almost "empty" though still just border on the human face. And they are also in an interesting way "readable" - as screens of projection.

Projections and Interpretations

In a Hollywood B-movie from the 50's, Claude Rains plays the role of a man without a face. It is the story of a scientist made invisible through the injection of a serum, and who as such commits numerous crimes: "The Invisible Man". In order to even exist for the others this "figure" must

wear clothing - no small challenge for the trick technology of the day. It was possible to read the most horrible things into this imaginary face. Behind the disguise of non-recognition lay all the potential of treachery, but also of possible romance. Such a "non-face" presents the ultimate stage of projection. Similarly the mechanical, decorative iron faces of the helmet visors may also be seen as examples of perfect projection.

The visors alter the true human face. They are artificial and - as shown by the preceding comparison with emotional typologies - are void of emotional and communicative associations. The visors are also a conscious effort to make one appear foreign, being created as they were during an epoch when the positions of "self" and "others" were being newly defined in social and philosophical terms. They are, however, timeless in their effect and continue, in many variations, to influence modes of expression to this day. On display today in the Zeughaus of Graz and presented in this book as a series of "portrait studies", they are removed from their original context - though they did actually function according to the psychological patterns outlined in this chapter - and are transformed into objects of general interest and influence. This is indeed only possible due to their disposition as objects of projection. Their innate character of estrangement is still visible today.

As such, the figures of heavily armed knights with closed visors have a startling resemblance to the visions of inhuman figures of the imaginary

future. Even early illustrations of robots in Karel Čapek's science-fiction phantasmagoria from 1920, in which mechanical androids appear for the first time, show an armored creature combining a mixture of insect-like and technicized features. Continuation of this line of thought gives birth to a creature without emotion and willpower, completely void of personality. The metal shell stands alone, there is nothing left which could hide behind the facade. On the other hand - and also typical of science-fiction - is the tendency to emphasize the inconsistency between the outer shell and the inner substance. Numerous martial villains of the genre - think of the figure of Darth Varder in Georg Lucas' film trilogy "Star Wars" who is squeezed into a mixture of morion, German military helmet and gas mask - are characterized by the frightful, depersonalizing aspect of a closed visor. Science fiction seems to delight in combining a physically weak inner figure with a starkly contrasting martial exterior. Thus we have a modern-day version of the split between the "self" and the "covering" inherent in the armor's function. As true with all modern tendencies, this version of splitting appears more drastic and irreparable. Where Tancredi's piercing of Clorinda's armor had a healing function despite the pain, where there was still something presumed to be concealed behind the armor, the new situation is different. In an age which uses science fiction to question the basic concept of personal identity, the answer could be quite different: behind the facade of armor lies decay.

And yet this fear does not hold true. As it turns out, the opening of the black mask is even for the popular mythical figure of Darth Vader a means of salvation. And even the mass media of film discovers the same thing that Tasso had presumed: beneath the steel is truly a human individual. One who feels and suffers, who has only been transformed by the armor to an evil, impersonal being. When the iron is broken it reveals - even in death - a human being.

In this respect, Karel Čapek's robots (after his futurist belief in progress was weakened) are reduced to little more than apotropaic figures, functioning in a manner similar to the suits of armor, as a picture of warning which wards off evil. Though the idea of depersonalization is deemed possible, the robot should serve as a warning. Look at what a monster you are creating. Be careful how you deal with increasing technology and alienation. Both robot and rider are steel personifications of warning and resistance. And both figures share another aspect which has linked them over the centuries - no matter how much the iron faces and steel men frighten, they never cease to fascinate. The fear and fascination of "facelessness" arise from the same source: the possibility of masking one's own identity with an alien and functional replacement.

For the history of edged weapons and armaments see
Peter Krenn, Swords and Spears, Landeszeughaus Graz, Ried 1997.
Peter Krenn (editor), Der Grazer Harnisch in der Türkenabwehr, Graz 1971.
My portrayal of the literary tradition of medieval knights is taken from
Joachim Bumke, Höfische Kultur, München 1986,
to whom I also owe the quotations regarding the banning of the longbow and crossbow.
About the history of the army and drilling: Michel Foucault, Discipline and Punish, 1977
Regarding the neo-platonic characters of universal-harmony see:
Thomas Höft, Die Machbarkeit der Welt, in Concerto Nr. 45, Köln 1989
Philippe Braunstein's theory about Emperor Frederick's face is quoted from:
Philippe Braunstein, Annäherungen an die Intimität;
Philippe Ariès and Georges Duby (editors), Geschichte des privaten Lebens, Band 2;
Frankfurt 1990.
Remarks about the biological foundation of physiognomy are based on:
Irenäus Eibl-Eibesfeldt, Die Biologie des Menschlichen Verhaltens, München 1984.
C.H. Hjortsjö, Man's Face and Mimic Language, Malmö 1969.
Siegfried Schmidt (editor), Der Diskurs des Radikalen Konstruktivismus; Frankfurt 1987.
On the apotropaic in the romantic period see:
Horst Bredekamp, Wallfahrt als Versuchung;
in Martin Warnke (editor), Kunstgeschichte. Aber wie?; Berlin 1989.
The episode of Tancredi and Clorinda from Torquato Tasso's *Jerusalem Delivered* is taken
from Judith Kates, *Tasso and Milton,* USA, 1983.

Daily Borders

April 1998
Looked for comparisons and found none.

All the helmets on the beams,
....birds' nests? beehives?
.... much too cool for this description.
Instead, the sight of rows:
only see the structure,
don't notice form.
And walk closer:
glance along the rifle stocks,
until they blurr
and only lines remain, vertical stripes
against the light.
And when the sun shines in,
a thousand bright reflections
glitter from the polished arms.
Mid-summer day in metal.
The stands of elaborate armor,
something between theater and factory.
Steel theater,
yet no one would imagine
that the armor comes alive,
marches forward, drawing swords...
All is calm,
Even the sun's reflections
don't enliven.
They structure.

A Question of Distance

At the time when mass usage of firearms becomes crucial to military success and death begins to recede into the distance, another instrument appears which brings distance closer for the individual. Around 1600 Galileo Galilei uses a telescope to observe heavenly bodies (a few years earlier, completely independently, this optical instrument had been used for the first time in Holland). The telescope verifies what the philosopher and scientist Giordano Bruno had metaphorically maintained and propagated: Bruno's observations led him to believe in the existence of innumerable other solar systems in addition to our own. A colossal discovery which revolutionizes not only astronomy. Infinity, according to Bruno, is ultimately to be found in the countless rays of the sun, which upon striking the eye of the observer, would tear him apart in a kind of ecstatic act. Galileo's telescope bundles the rays of light and returns them to the viewer's eye allowing one to see distance up close. Although the enormity of this revelation does not destroy the viewer physically, it is indeed an overwhelming thought. For a human being who believes in the earth's central position, this is similar to being flung by centrifugal force from the center to the outer edges of the universe. As these new ideas approach the limits of human comprehension, the watchmen of the inquisition step in. Giordano Bruno is burned at the stake in Venice, and Galileo is forced to revoke his theories of natural philosophy.

This revolution in values that shakes the foundations of the old Christian-European world around 1600 is best visualized by two extremes: the overcoming of distance and speed. Mass killing from a distance and the possibility of covering unimaginable stretches in a moment are two aspects of the same development. If one adds a third aspect, the possibility of exactly measuring minutes and seconds (a discovery of the watchmaker Jost Burgi at Emperor Rudolf II's court in Prague) then the tremendous shaping influence of mechanization becomes obvious. Instruments divide the passing of time into small, infinitely equal parts, like a mechanical heartbeat which relentlessly segments day and night. Instruments bring foreign worlds, like the moon, close up, almost within reach. And instruments are gradually enabling one to kill one another without approaching too closely and putting oneself into danger. All this is not yet perfected, but the direction of the course to be followed - which has parallels to today - has been chosen. Humans must get used to other distances, acceleration has begun.

Governmentalization of War

Michel Foucault describes the fundamental change in warfare which occurs during this period as follows: "Military operations and institutions have become more and more concentrated in the hands of a central authority, so that only government powers can defacto and de jure instigate war and implement military equipment. Thus war is governmentalized. Consequently, one kind of social body, a form of inter-personal relationship, the so-called daily war or private war, has disappeared. More and more often wars take place only on the border, confined to the outer

regions of large political bodies - as a possible threat or real measure of power between these states ... Because of this governmentalization which pushes war to the outer boundaries, war becomes the professional and technical monopoly of a carefully defined and controlled military apparatus. This is the beginning of the army, which did not exist as such in the Middle Ages ... It is paradoxical that during and after this transformation, which on the one hand centralized the war, and on the other hand, pushed it to the state's outer borders, a certain discussion surfaced, a novel type of discussion ... And what does this discourse propose? Contrary to philosophical and juridical theories, it maintains: political power does not begin when war ends. The legal organization of power, the structure of governments, of monarchies, the principles of society are not found there where the weapons have been silenced. The war is not over. First it gives birth to the state. Principles emerge from burning cities and devastated countries ..."

How political power emerges from devastation is best demonstrated by the situation on the military frontier. Although the Croatian nobility refuse already in 1558 to agree with the destruction of fortifications east of the newly proposed borderline, their protest is to no avail. In 1578 Emperor Rudolf II officially transfers responsibility for defending the border to

Match-lock Guns

Archduke Karl, the sovereign of Styria, Carinthia and Carniola. This frontier region turns into a permanent battleground, especially once the main defense troops settle in. For decades, it is not the big wars but rather the small, relentless attacks, the destruction, ransacking and murdering on a local scale which characterize the conflict. War in the border zone becomes a daily occurrence - devastation and slaughter, the rule.

Taxation and Forced Labor

The most important instrument of power possessed by the estates was their right of taxation. Although the archduke of each province could claim a budget for defense purposes, he had no legal authority. Especial-

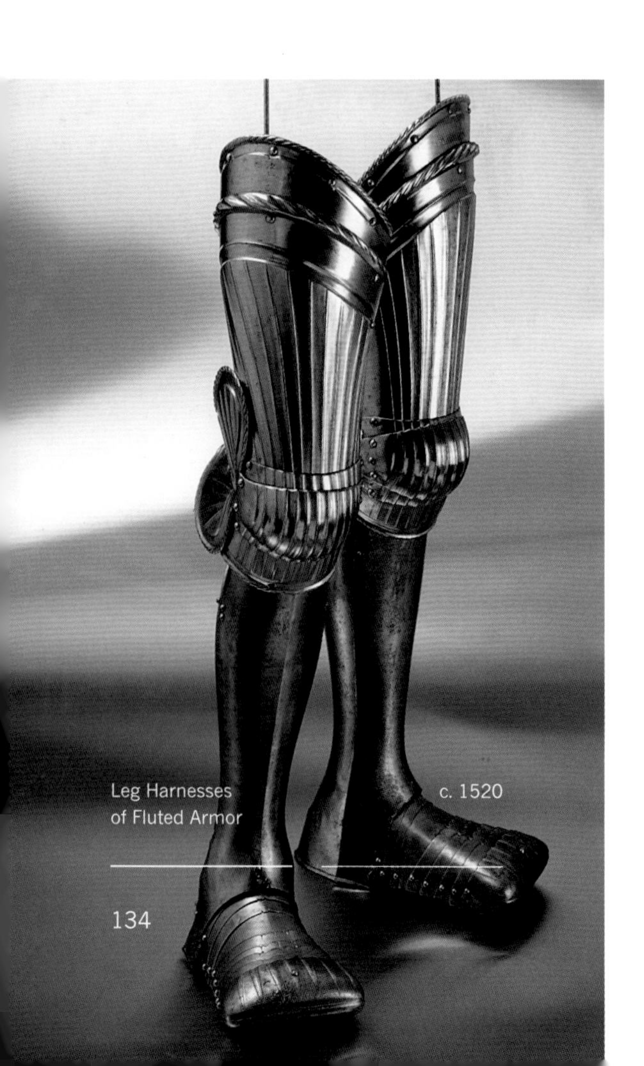

Leg Harnesses of Fluted Armor c. 1520

Fluted Armor
Hans Maystetter

Innsbruck 1510-11

ly with regard to the amount of funds he had to rely on cooperation with the estates who, in addition to their basic legal authority, also determined the form and extent of taxation in the Provincial Diet. The tax system was based on the so-called "Gült", or ground rent. This ground rent was imposed by property owners according to the interest profits of their subjects as well as income from further rights such as mining rights, etc. Landowners, on the other hand, levied taxes measured on the amount of interest the vassals earned.

This dense jungle of permissions and concessions, of payments in kind and service, of performance and financial rates of payment is difficult to comprehend and cannot be handled in this context. The most important fact to remember is that taxes from 1543 to 1624 increased to four times the ground-rent and that additional duties such as personal, building and property taxes added further hardships. Since the landowners collected these payments directly from their vassals it became an almost unbearable burden for the "common man", up to as much as ten times the ground rent.

A further instrument of power employed by the estates, which increased military preparedness while putting additional pressure on the population, was the compulsory work force. In case of a threatening situation, the estates were authorized to recruit men for work such as constructing fortifications or setting up barricades. What constituted a "threatening situation" was defined by the estates themselves, and at times this compulsory work took on dimensions which made it impossible for the men to carry out their normal agricultural duties. Forced labor posed an acute threat to the livelihood of the peasantry. The impoverishment of entire regions was due not only to constant military attacks and overwhelming taxation, but also to the compulsory work measures which prevented any economic progress. The peasants' once proud idea of "helping themselves" is transformed in the 16th century to a firm dependency on the ruling class, making them once again objects in the hands of the powerful. The only chance to regain any self-sufficiency was for farmers to resettle on the border to Croatia. There the conditions were more dangerous but also provided more freedom.

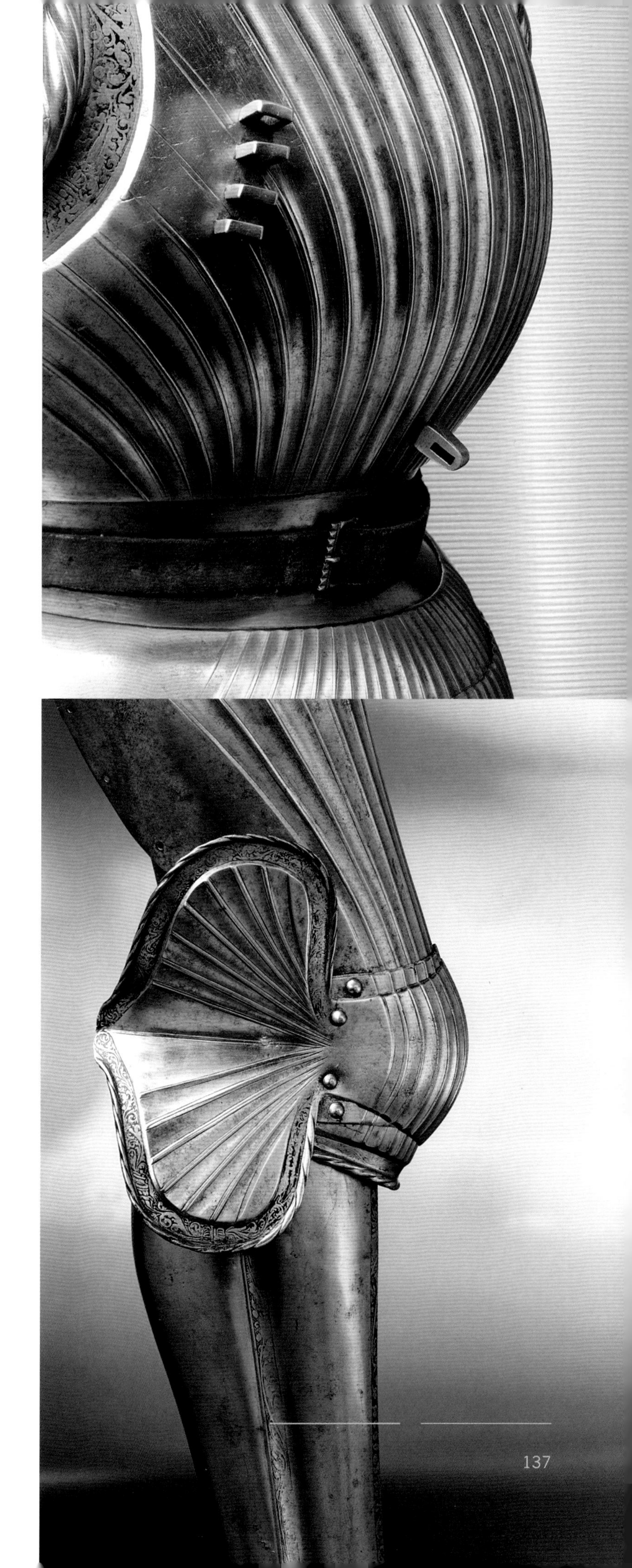

Frontier Settlers

The military frontier was organized in four sectors and administered by the Imperial War Ministry *(Hofkriegsrat)* in Graz, thus under the rulership of the provincial sovereign. The financial responsibility for supporting this frontier was assumed at the previously mentioned Provincial Diet of Bruck an der Mur in 1578, thus by the estates of the provinces of Styria, Carinthia and Carniola.

Large areas of this border zone had become no-man's land. Incessant plundering had decimated the local population and finally driven the rest away. Farmsteads were largely destroyed, the lands laid waste. In order to secure the border, it was necessary to re-populate these deserted regions. Yet how does one convince peasants to resettle in a war zone which offers absolutely no guarantee of a safe life? The answer lay in far-reaching privileges offered to all prospective settlers.

Resettlement of the border region is not a one-sided measure. One should not overlook the fact that just as the western, Christian side was preparing to secure the status quo, identical efforts were being carried out on the Turkish side. The most important supporters of the military frontier on the western side were the Wallachians, families of Serbian Orthodox faith, who switched over from the Turkish side to western territory (a new humiliation for the Croatian population, who after losing their sovereignty, had been defenceless in the face of the Turks, and who now were being replaced and driven away by a population from "the other side"). The Wallachians received extensive concessions. In exchange for their pledge of constant military preparedness they obtained their own land as free farmers. Although former landowners were initally unwilling to relinquish their feudal rights by which the settlers would become vassals, the military administration dispersed all claims by convincing them of the necessity of border defense in their own interest . Finally in 1630 Emperor Ferdinand II made these Wallachian privileges, the "Statuta Valachorum", official by decree. As such the farmers were free landowners, could make independent decisions about their property, and were freed of all taxes. The pieces of land did not have to be acquired, but were bestowed on the new settlers. And even the local jurisdiction lay not under aristocratic control, but directly in the hands of the village inhabitants who elected a village judge for this purpose once a year.

These new settlers were no longer defending an abstract state, but their own personal property. Of course, there were other problems in addition to the constant threats. For instance, land in the district of Varaždin was much more fertile than in the generalship of Karlstadt where famine repeatedly broke out. But on the whole, the situation of the farmers on the border with regard to personal responsibility, independence and authority was quite contrary to the development in the imperial territories. Displacement of war to the border was the strategy of a modern-day political system which no longer distributed privileges and subordination according to religious or genealogical justification, but purely according to expediency. Freedom there where it serves the system, repression and exploitation there where a clear stronghold of power is guaranteed.

Breastplate of Fluted Armor c. 1520

War Profits

Armament is a business. Freedom on the border - justified by the ever-present fear of "the other", yielded maximum profits. The number of scandals - subsidies which disappeared into unfathomable channels, construction which was unprofessional and ridiculously expensive - multiplied.

One example of the entanglement of power and war profiteering which was possible at the time, can be seen in the fortress construction in Kopreinitz. In 1156 it is decided that the badly damaged medieval fortress should be renovated and modernized. Vassals of the Croation estates are obliged to carry out the construction, the Styrian estates provide the building material as well as a large financial contribution to aid payments. Construction continues until 1574, but when the engineer Alexander de Pasqualini inspects the almost completed fortress, he has to ascertain that the intended plans have not been carried out. The costs explode due to ridiculous changes and the final result is far from complying with any "rules of architecture".

Whereas the border administration at times had little more to do than distribute subsidies, the living situation of the frontier settlers remained, in spite of all social freedom, precarious. Although it was not considered real warfare, both sides tried to weaken the other by surprise attacks, the so-called "Tschettieren" (derived from the Croatian word "ceta" which means "small troop"). These frequent incursions of small armed bands were aimed at destroying new settlements, ruining crops and murdering or abducting the inhabitants. There was no difference in the cruelty perpetrated on both sides. A Christian chronic from 1575 reports: "On January 22nd, Turks attacked our midst and kidnapped 500 people. In February, Turks stormed the fortress of St. George. In May, large number of Turks attacked St. George, set fire to surrounding villages, 50 persons, 60 horses and 400 animals carried off. On June 12th, large number of Turks stormed the fort of St. George, two churches burned. In August, Turks attacked Copreinitz in large numbers. In same month, outskirts of Copreinitz set on fire." This list of terror is certainly useful as propaganda against the Turks and to convince of the need for higher taxes. If, however, one reads the accounts written in 1562 by Veit von Hallegg, a Styrian first

Leg Harness
of Fluted Armor

c. 1520

Field Armor of Archduke Karl II.

Conrad Richter
Augsburg 1565

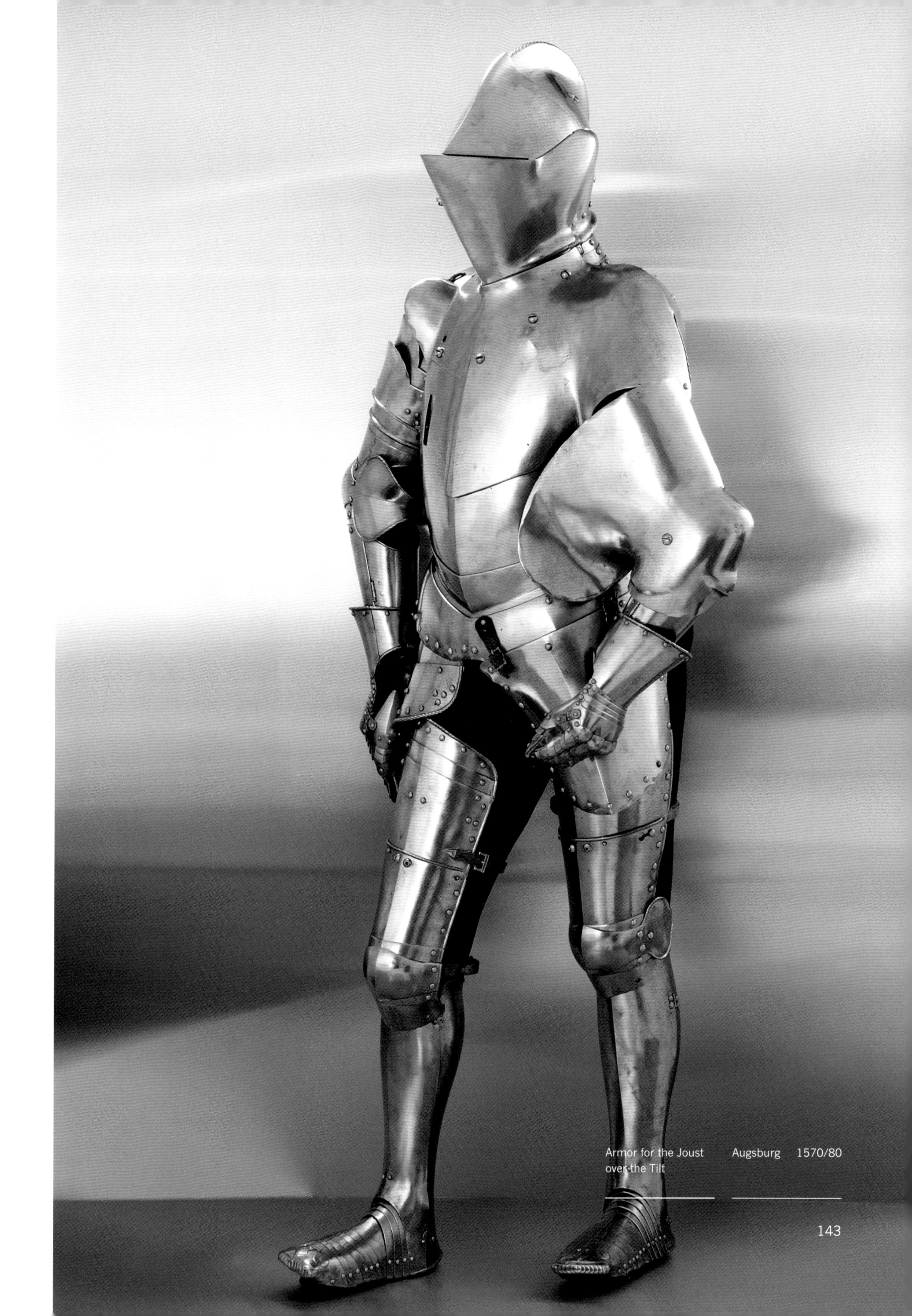

Armor for the Joust over the Tilt | Augsburg | 1570/80

Garniture for Tournament of Kaspar Baron Völs-Schenkenberg

Michel Witz the Younger
Innsbruck 1560

Garniture for Tournament of Kaspar Baron Völs-Schenkenberg

Michel Witz the Younger
Innsbruck 1560

lieutenant on the border, the events appear in a different, even more horrible light: "The raid went well. Destroyed a large number of villages, set fire to more than 1,000 houses. Also completely burned towns around the fortress of Slatenay. There were 370 soldiers at the fort and many people who fled inside for protection. They resisted for quite awhile, but finally we set the castle aflame. No one survived the fire."

There is no good or bad side per se in this continual war. The perversity generated by violence is universal and not a question of religion. Another example of this is the flourishing trade with children. The Ottomans used abducted children not only to form their elite troops, but also for their well organized civil service. Free of all family ties and thus of any possible personal interests, these children are raised as strict Muslims and later become the most loyal civil servants. The demand for children was fulfilled not only by abduction during military raids, but in addition, by border inhabitants on the western side who sold children to the Ottomans. Slave-trading in general became a profitable business. Mutual raiding on both sides often yielded a considerable number of persons as bounty who could be sold or traded for other valuables; a common practice accepted by all parties. The border was much more of a membrane than an impermeable line. Trade routes crossed it along with military movements; indeed, several fortresses became regular trade centers and experienced a period of steady economic growth during this period.

Arms Boom

The challenge of defending the outlying military frontier while at the same time maintaining independent provincial troops, led from 1575 to a massive increase in armament. The Styrian estates, financially responsible for both border and army, placed large orders for arms thus causing a boom in the weapons industry. The necessary material and know-how,

Parade Armor

Michel Witz
the Younger
Innsbruck c. 1550

Horse Armor of Georg v. Stubenberg-Wurmberg

School of Conrad Seusenhofer Innsbruck c. 1510

however, had to be purchased from the German imperial territories, since the local capacity was by no means sufficient. The main center of arms production at the time was the free imperial city of Nuremberg. In 1575 the Styrian estates dispatched a request to Nuremberg's City Council asking if they could send a few experienced master-gunners to Graz - the request was promptly fulfilled. In addition, large quantities of high quality gunpowder were ordered from the Nuremberg dealer, George Sonner. Two years later Sonner also received his first substantial order for arms: within several months he delivered more than 2,500 marksman's helmets, 180 pieces of light calvary armor, and 180 pieces of armor for the "Landsknecht", 1,169 wheel-lock guns and 57 wheel-lock pistols including all necessary accessories. This was the beginning of further large shipments from Nuremberg to Graz which, over the next ten years, totaled approximately 3,000 pieces of armor, over a thousand guns and almost the same number of pistols. In addition to the estates, Styria's reigning sovereign also ordered weapons so that Graz was "probably one of the best armed European cities of the day" according to Peter Krenn and his study of arms deliveries from Nuremberg.

The sovereign's weapons were stored in a newly erected armory in the Hofgasse as well as in the Schloßberg (castle) of Graz, whereas the estates divided their weapons between arms depots near the city gates and the attic of the 1565 newly completed Landhaus on Herrengasse. This ever increasing amount of armor and weapons soon made storage space a problem, and discussion of how to insure a sensible and professional storage of the equipment began.

Amassing Wealth

Meanwhile, the fact that the Styrian estates were responsible for financing the military frontier led many Styrian aristocrats to begin their careers in Croatia. Duty on the border was obviously still quite desirable despite the low ground salary. For one reason, not all peasants possessed special privileges - there were still many, especially in the big fortified cities, who were not free and as such were required to pay taxes to the fortress commanders. In addition, these vassals could also be conscripted for general

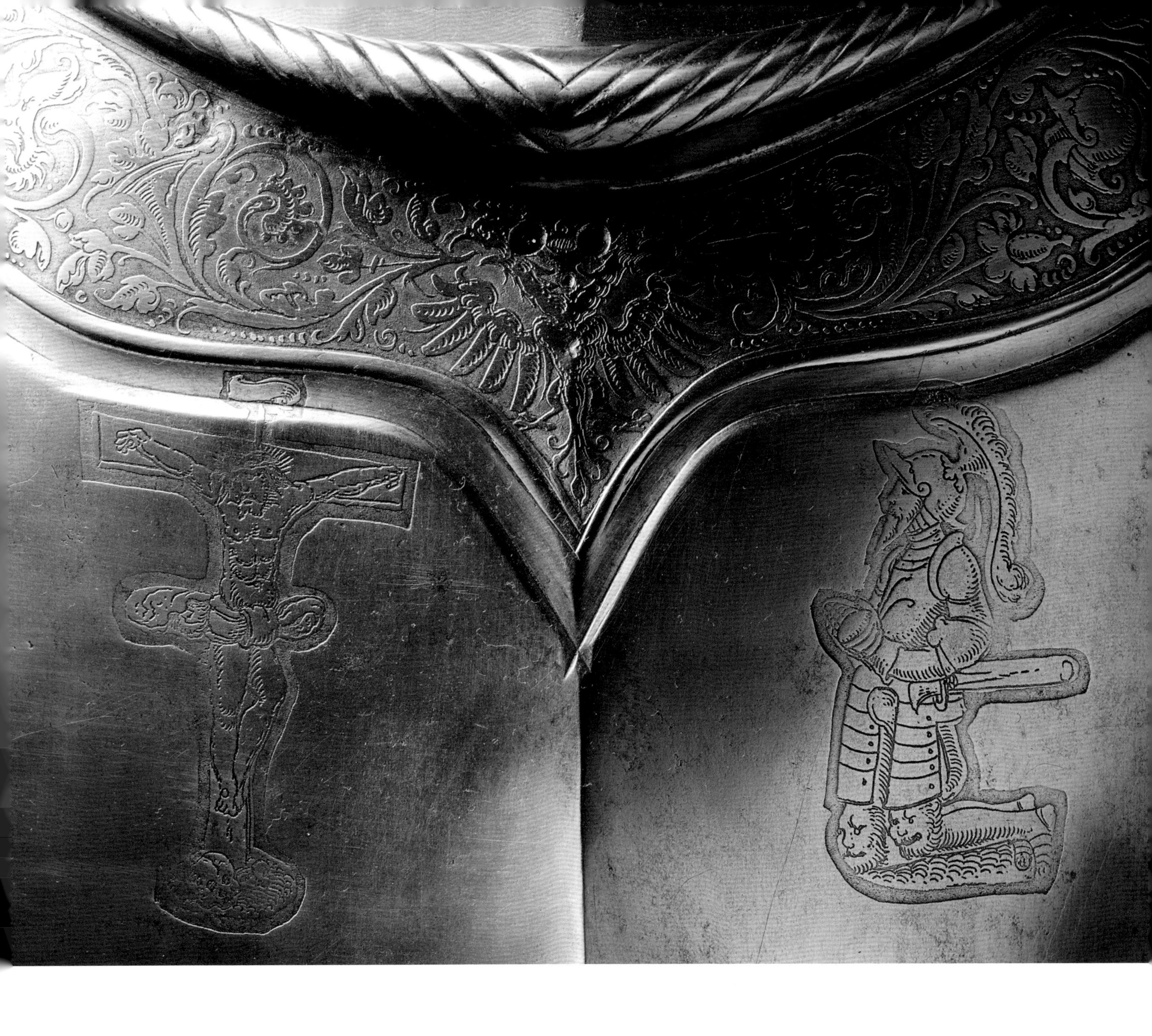

Etched Breastplate of Field Armor

Mid 16th century

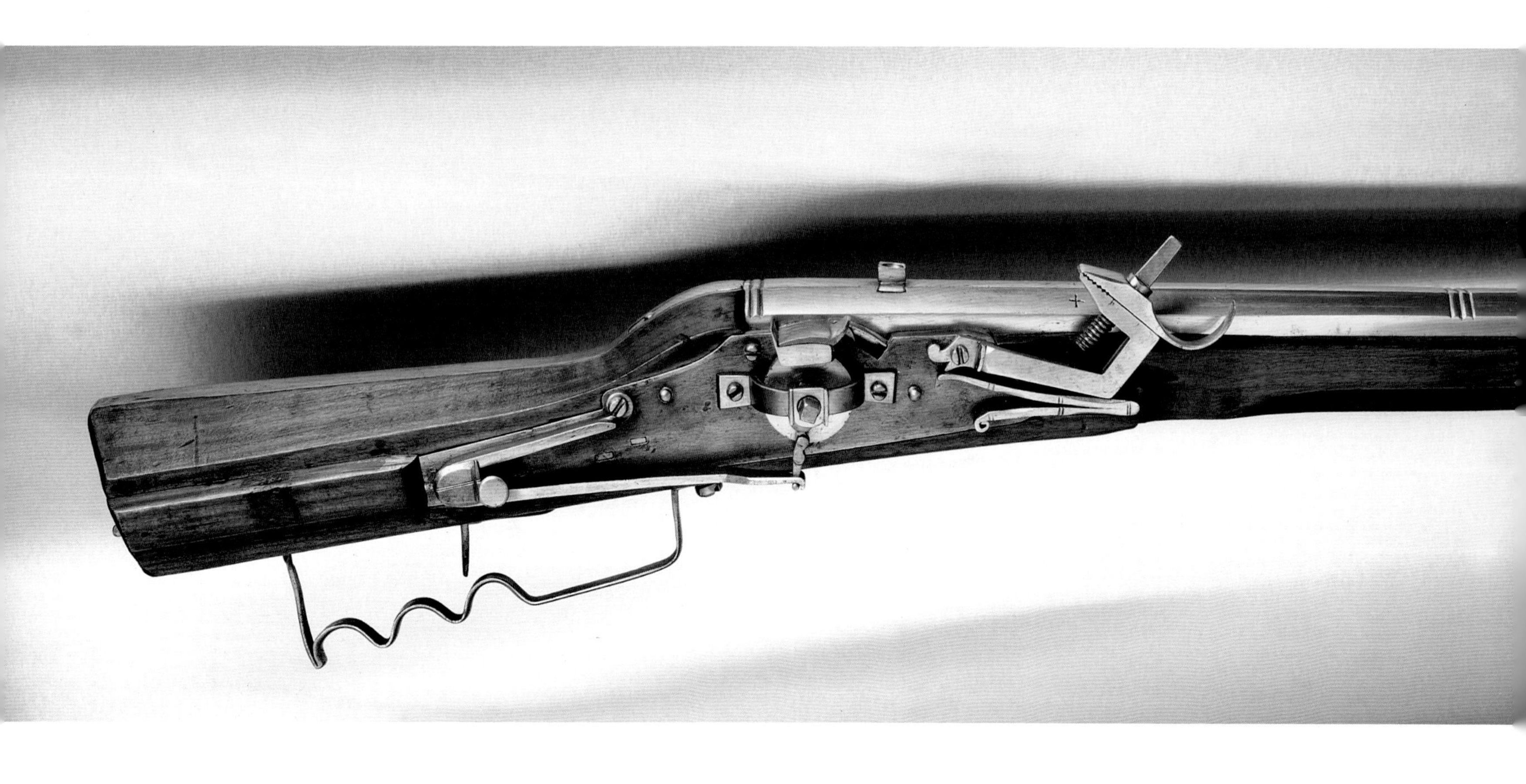

services. Thus it was possible for the captain of Creuz to collect, in addition to his base pay, the revenue from 90 "yokes" of arable land, grassland, forest and vineyards which had been cultivated by serfs. Besides this basic "means of existence", there was always the additional chance of gaining real wealth through war spoils. The constant raids on Turkish territory were a lucrative source of income , especially since the soldiers were required to divide the bounty with their captains. Turkish valuables, such as rugs and jewelry, piled up in the fortresses. Just as desirable were the prisoners which were either resold or taken as personal slaves.

Pillaging and enslavement were part of a situation which lacked almost any semblance of law and order, where only power and pure violence counted. Not before the turn of the 17th to 18th century, when command of the border was actually taken over by central authorities in Vienna, was plundering forbidden and a governmental system of law effectively spread and reinforced throughout the border districts.

Till then, however, petty dealing blossomed, war material was transferred and sold at personal profit, the serfs were mercilessly exploited and the Turkish side was relentlessly ransacked and looted. The lawless situation on the border even led to an economic crisis, as described by Helfried Valentinitsch: "From 1620 due to a lack of cash, the Turks offered to pay salt for the release of their prisoners. Soon Styrian officers had amassed such large supplies of salt that they could no longer sell it to the Croati-

Wheel-lock Gun	l	113,5 cm
Late 16th century	w	3,85 kg
	cal	14,3 mm

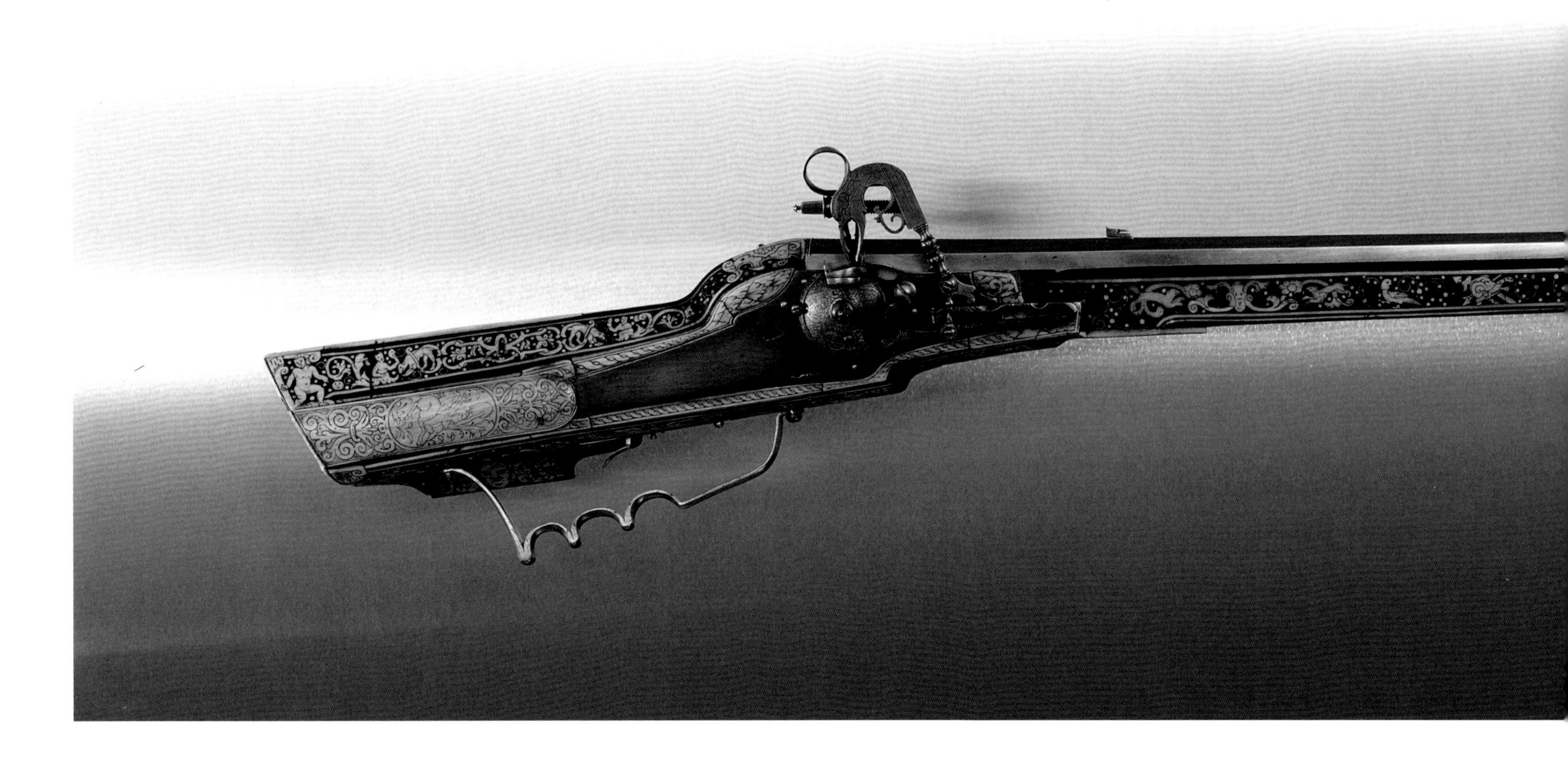

ans. Instead they smuggled it into Lower Styria, evading the imperial salt monopoly, and sold it here at an extremely low price. This smuggling of salt was facilitated by the fact that several times during the Thirty Years' War the Emperor had drastically raised the Styrian salt price. The result was a considerable disruption in the salt market of Aussee in Lower Styria." The history of Styrian domination in this border region is actually the history of a shadow economy with all its advantages and disadvantages. Though the displacement of war to the outlying regions partially calmed the situation in the inner territories, it also served to divert all manner of extremes to the outer membrane. As such the border offered both threat and enticement, a crossroads where the perversities of the day, the contradictions and problems became apparent. A festering wound and an unparalleled source of contagion, a testing ground for the brutal rules of conduct which would later govern the entire system.

Counter-Reformation

Whereas the - mainly Protestant - Styrian estates continued to feel secure in their power and maintained a self-assured position with regard to the provincial sovereign due to written privileges, a gradual change in the overall political situation began, leading toward a new crisis. Emperor

Wheel-lock Rifle	l	117,4 cm
c. 1600	w	3,95 kg
	cal	15,1mm

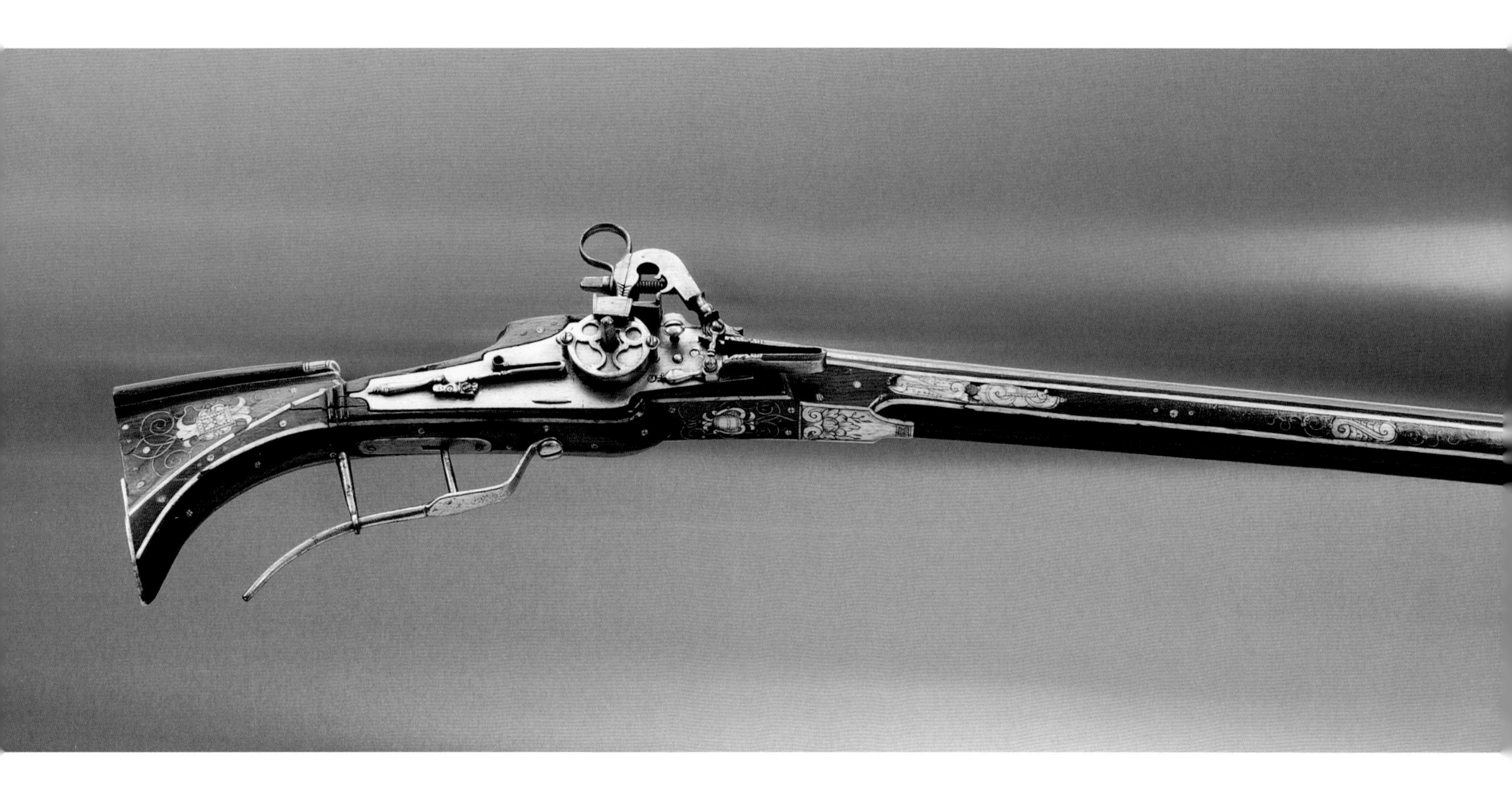

Rudolf II, though raised in Spain and convinced of the superior and sacred position of the Catholic faith, nonetheless conferred on the estates new freedom for political and philosophical reasons. Jewish theologists, Protestant scientists and Islamic scholars were called to his court in Prague in order to live the utopia of an intellectually perfect world society, this under the patronance of the Emperor. The reality, however, looked quite different. Confessional differences continued and grew into political rivalries, largely due to the Treaty of Augsburg with its linkage of religion and power. During this period of internal tension and strained relations, the Turks recovered from their set-back after the death of Suleiman, the Magnificent, and aspired to improve their military position on the Windisch Military Front. Their target was the strategically important fortress Sissek on the Save River.

In 1590 Archduke Karl of Inner Austria died. The Styrian estates repeatedly refused to pledge their oath of fealty to his successor, Archduke Ernest. The Provincial Diet also blocked new tax measures - without the renewed assurance of religious freedom the estates decided on a boycott. Meanwhile, however, the estates were partially responsible for the desolate situation on the border, where mismanagement and personal profiteering had already seriously endangered military effectiveness. When Pasha Hasan overwhelmingly defeated Styrian troops in 1592 at Sissek, those in positions of responsibility finally became aware of the seriousness of the situa-

Wheel-lock Gun
1600–1625

l	103,5 cm
w	1,77 kg
cal	13,4 mm

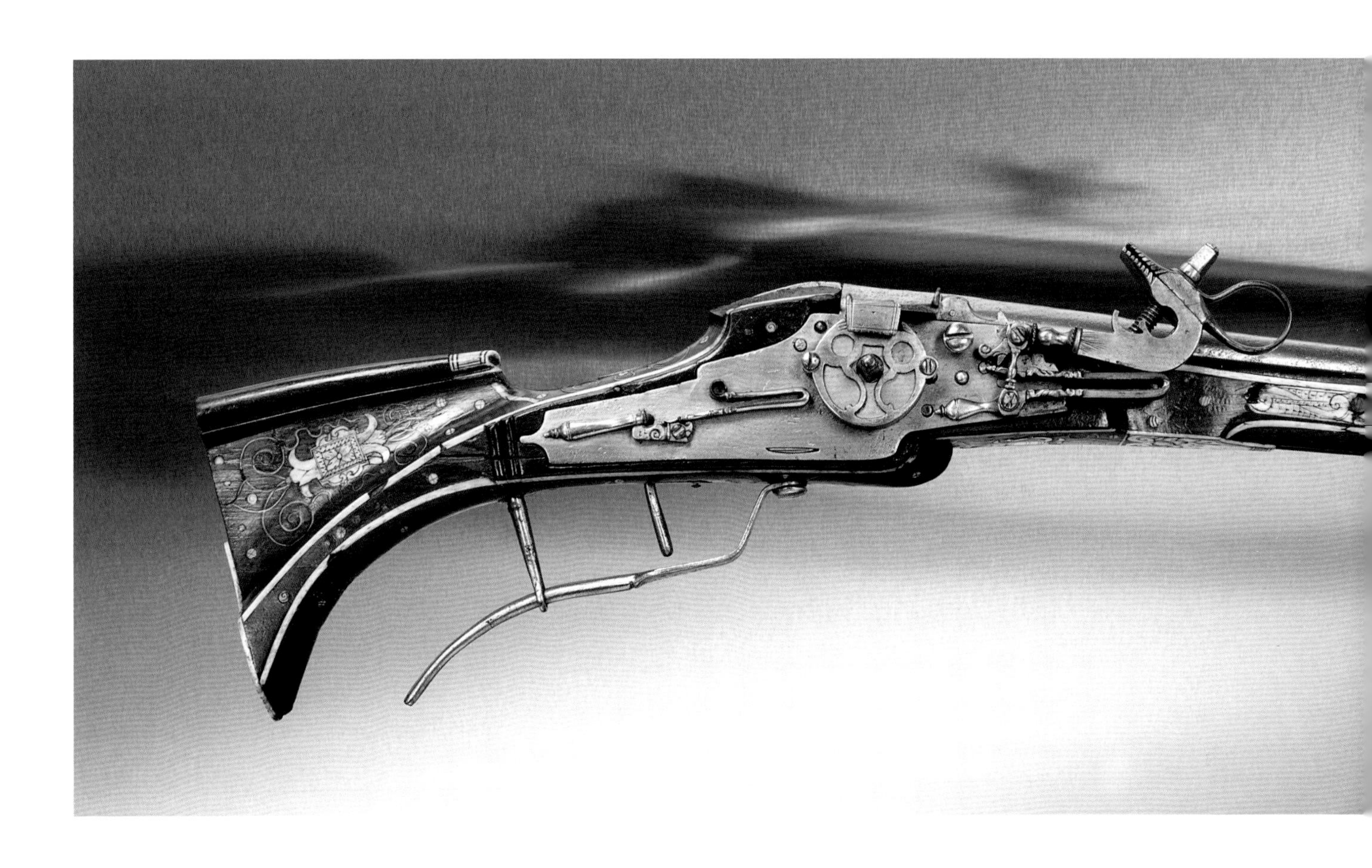

tion. The estates ended their boycott and accepted Ernest as representative of Ferdinand (who was still a minor) without receiving any new assurances with regard to religious freedom. The balance of power began to shift.

The "Long Turkish War"

In 1593 Pasha Hasan once again attacked Sissek with a large army. This time, however, he suffered a crushing defeat at the hands of an army under the leadership of Ruprecht von Eggenberg (providing an important basis for this family's meteoric rise during the Thirty Years' War). The Pasha himself was killed, and subsequently the Ottomans officially declared war on the Empire. During the following 13 years of bloodshed, fortresses were lost and regained, the Turks joined forces with the Venetians against the Emperor, and the latter allied himself with the Persians against the Turks - a decisive victory, however, was not achieved by either side.

When Archduke Ferdinand, the later Emperor Ferdinand II, officially took on his inherited position as provincial sovereign in Graz, the estates were confronted for the first time with a new kind of absolutist prince, who was uncompromising in all questions of faith and consequently political power. A year before, in 1596, he had visited the military border, which to a sub-

Wheel-lock Gun 1600–1625	l	103,5 cm
	w	1,77 kg
	cal	13,4 mm

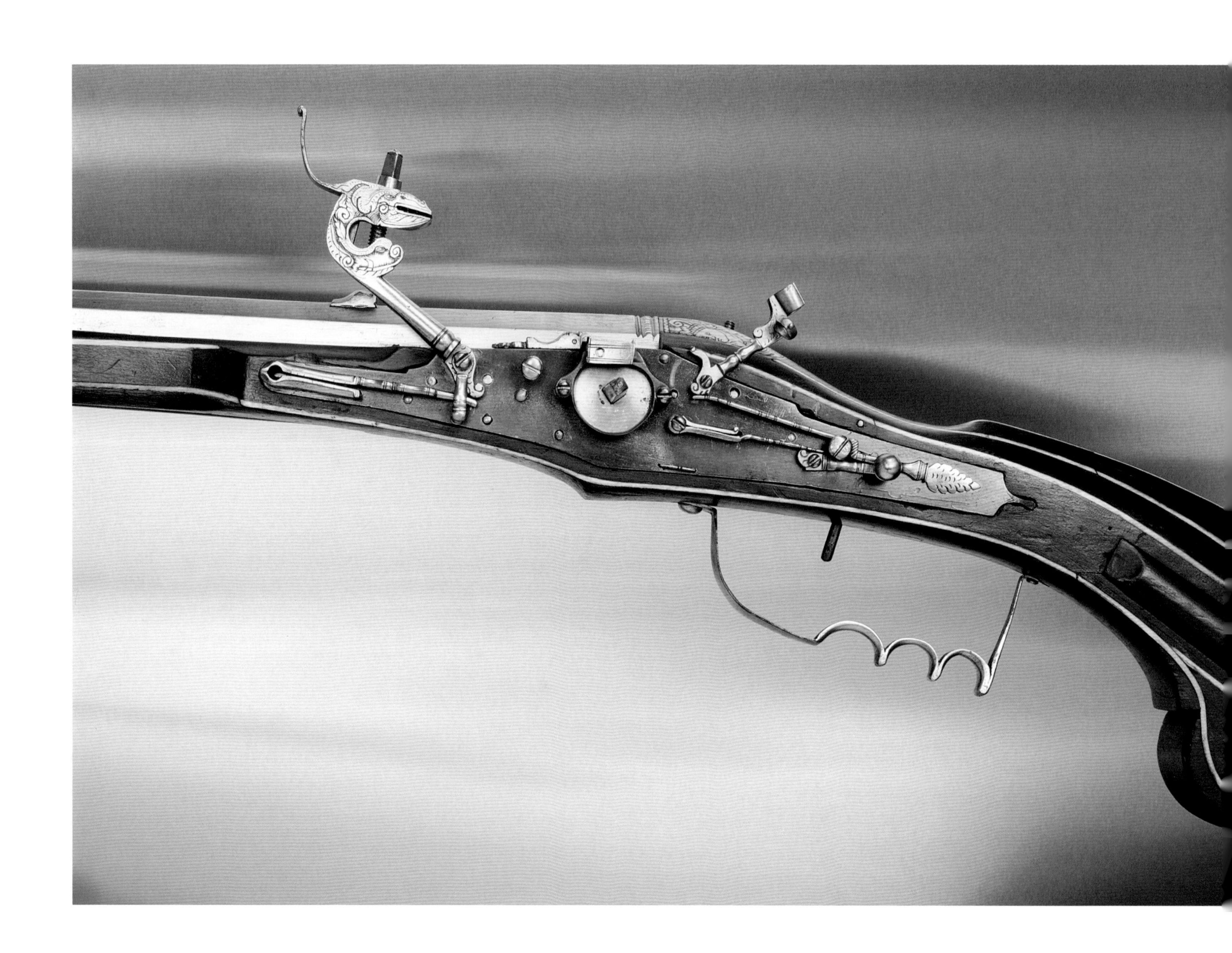

Wheel-lock Gun		l	138,4 cm
Augsburg	c. 1600	w	5,95 kg
		cal	17.3 mm

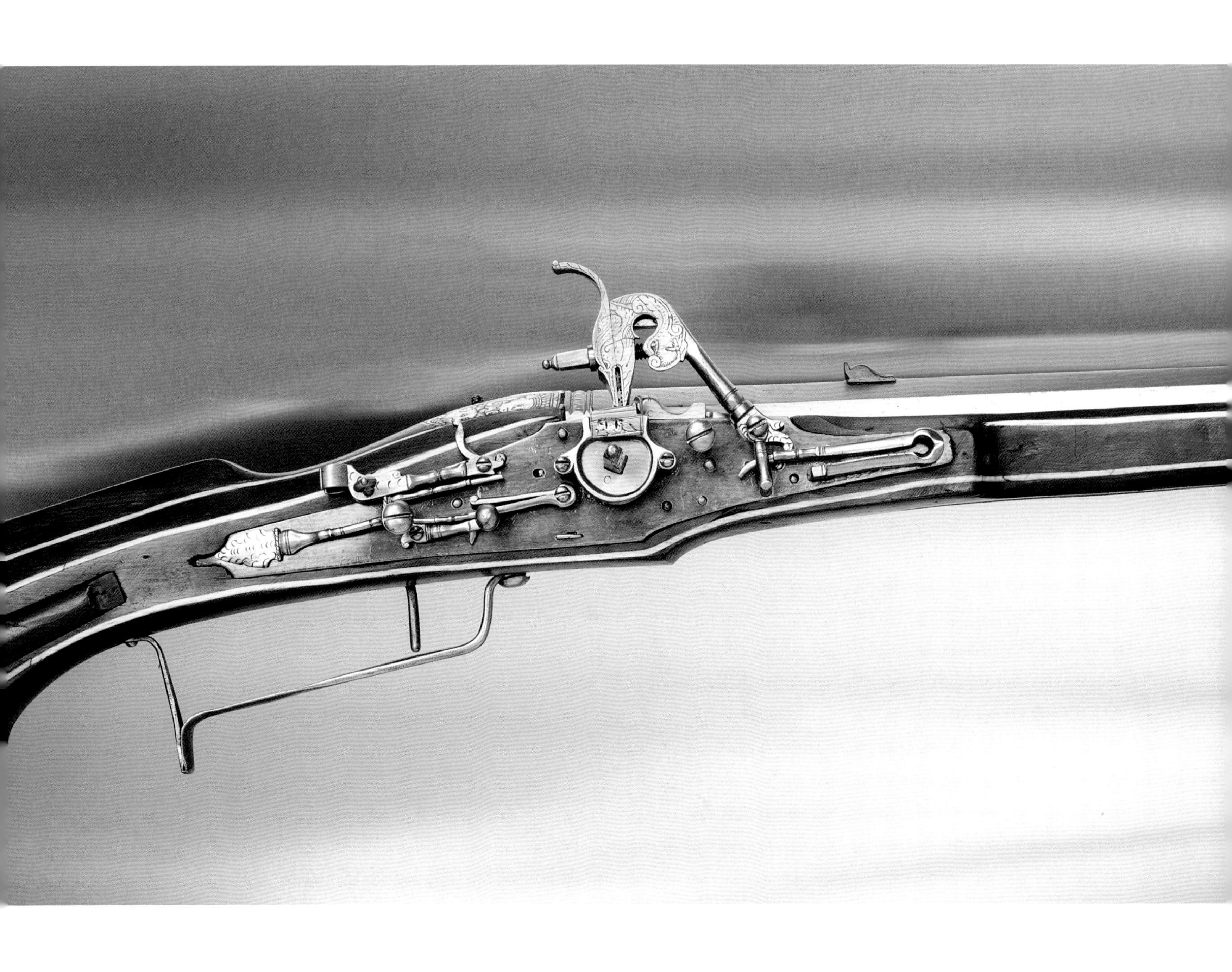

Wheel-lock Gun		l	141 cm
Augsburg	c. 1600	w	5,78 kg
		cal	17,3 mm

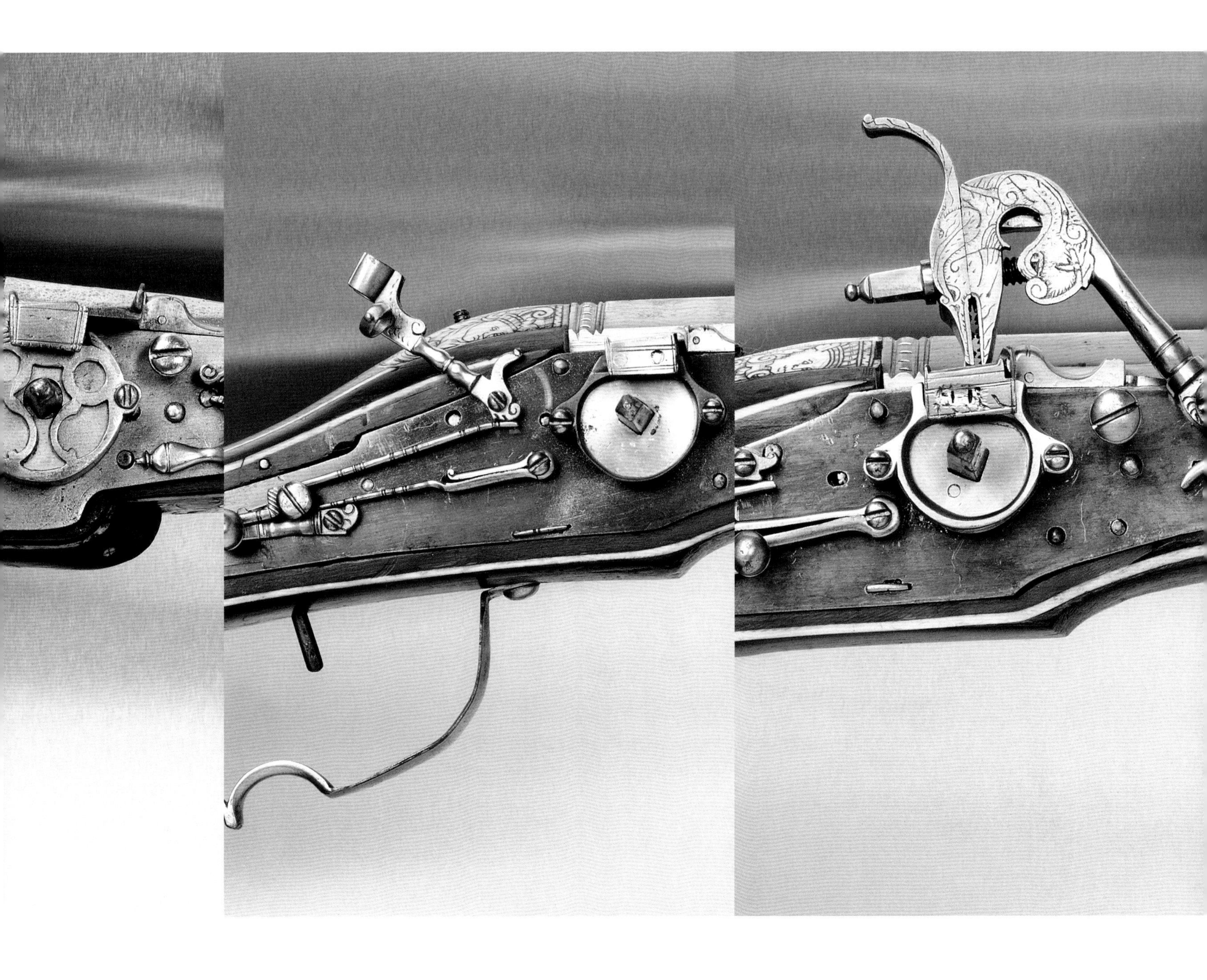

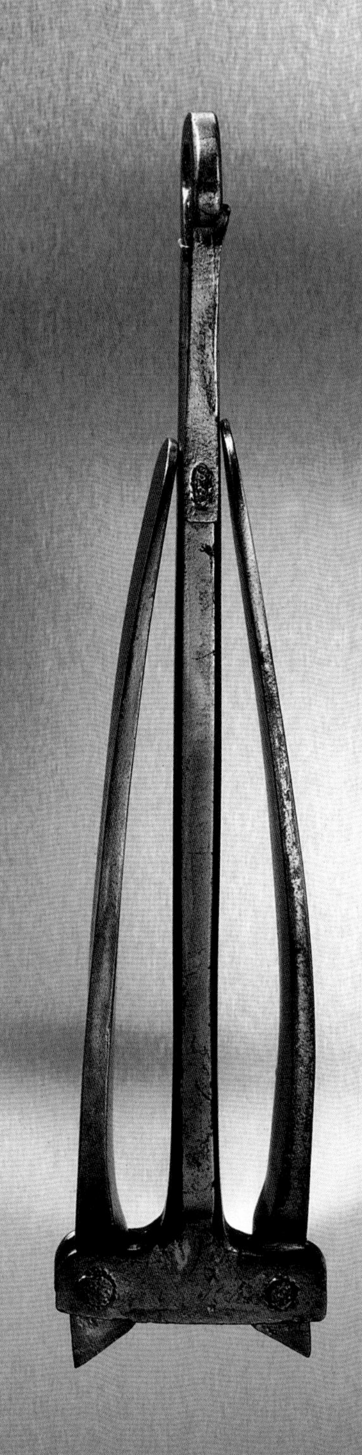

Collapsible Winch

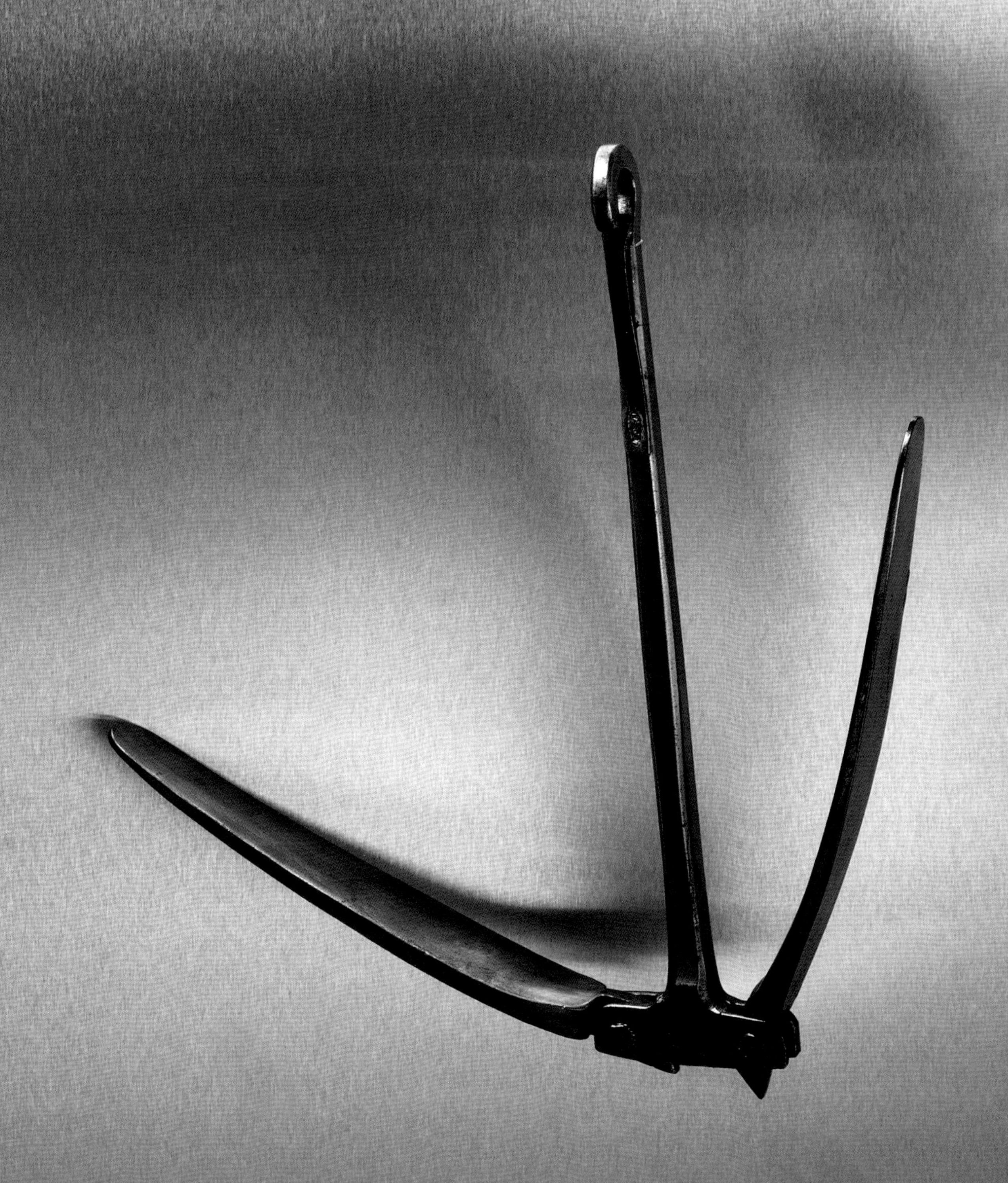

stantial extent was also financed by the imperial treasury in addition to provincial funds, had become convinced of the desolate situation and decided to take control of all further measures. In a scathing report to Emperor Rudolf he criticized the fortress of Karlstadt as resembling a "pile of dirt", and added furthermore that " almighty God only knows how such an open, uncontrolled and ragged border has ever managed to hold".

In his negotiations with the estates regarding terms of allegiance, Ferdinand, who had been strongly influenced by Jesuits since his youth (and who was "of a weak constitution and a bit dumb" according to his uncle, Archduke Ferdinand of the Tyrol) was not willing to make any further concessions. In 1598 he had all Evangelical preachers evicted from the province's cities and towns. The estates countered by refusing new tax allocations, a tool which proved ineffective because of dissent among the nobility. Little did they know that they would soon be brutally chased out of the country (or forced to convert).

In this situation imperial troops scored an important victory over the Ottomans, both in political and propagandistic terms. The fortress of Raab, which had fallen 1594 into Turkish hands, was once again recaptured with considerable booty. Backed by this triumphal success, Ferdinand proceeded to brutally carry out the Counter-Reformation. Citizens were forced to convert or relinquish their property and leave the country; serfs had no choice at all. Little by little, the Protestant aristocracy lost all influence (a development which was concluded by around the beginning of the Thirty Years' War). A wave of conversion set in, and in an incredible switch of position many old families lost their property while young families were quickly ennobled and elevated to fill these positions and functions. It was a true "revolution from above" to which families like the Eggenbergs and the Attems owe their advancement.

In spite of the victory of Raab, the war was far from being won. For instance, only 50 kilometers from the Styrian border, the fortress of Kanisza fell to the Turks. (It was rumored that the fort commander, George Paradeiser, deliberately played into Turkish hands in revenge for Ferdinand's counter-reformatory measures. Paradeiser was then executed.) An attempt to recapture the fort with an army of 27,000 men failed when winter set in.

Collapsible Winch

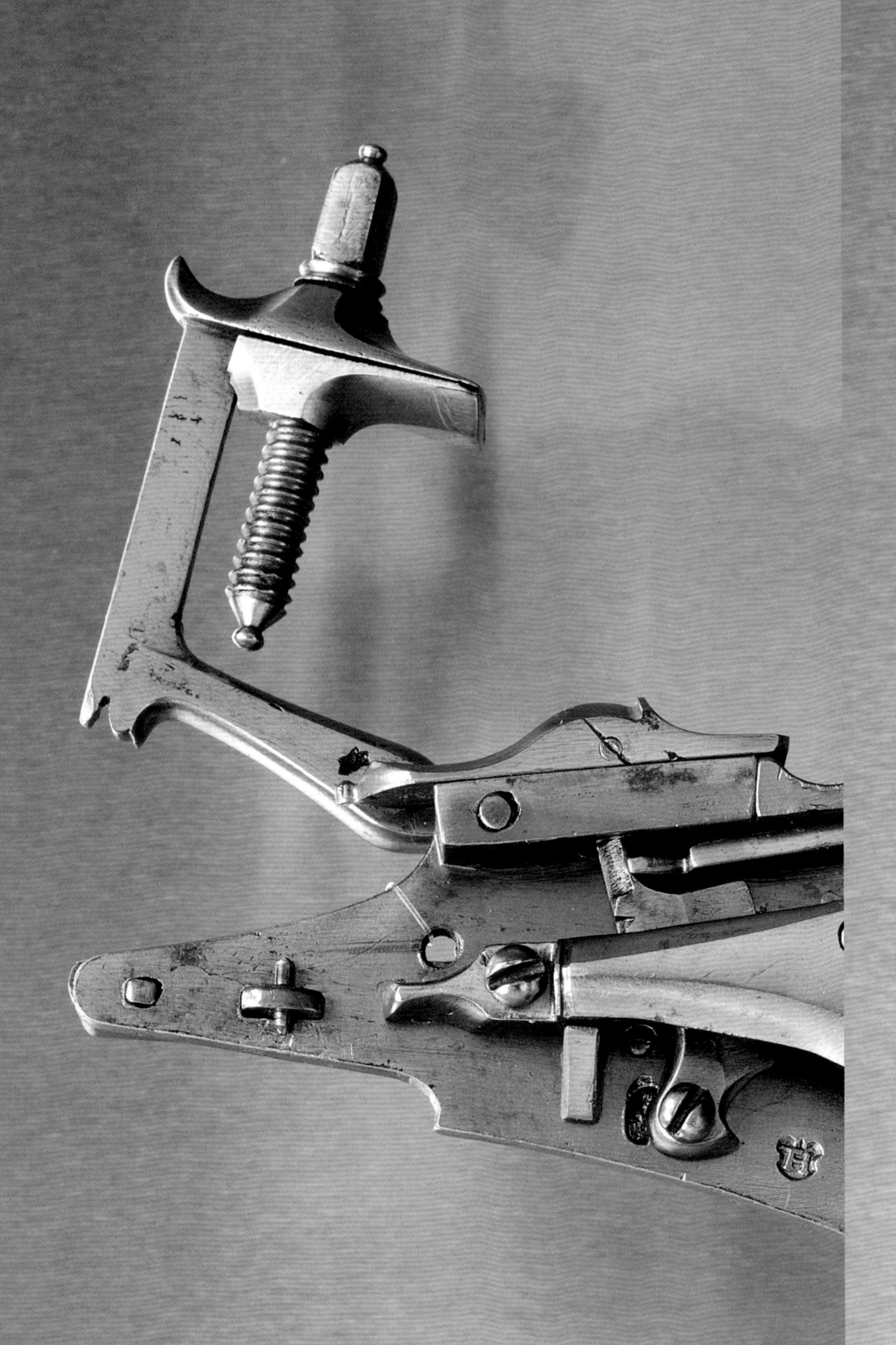

Wheel-lock Pistol

Ferlach
First half of the
17th century

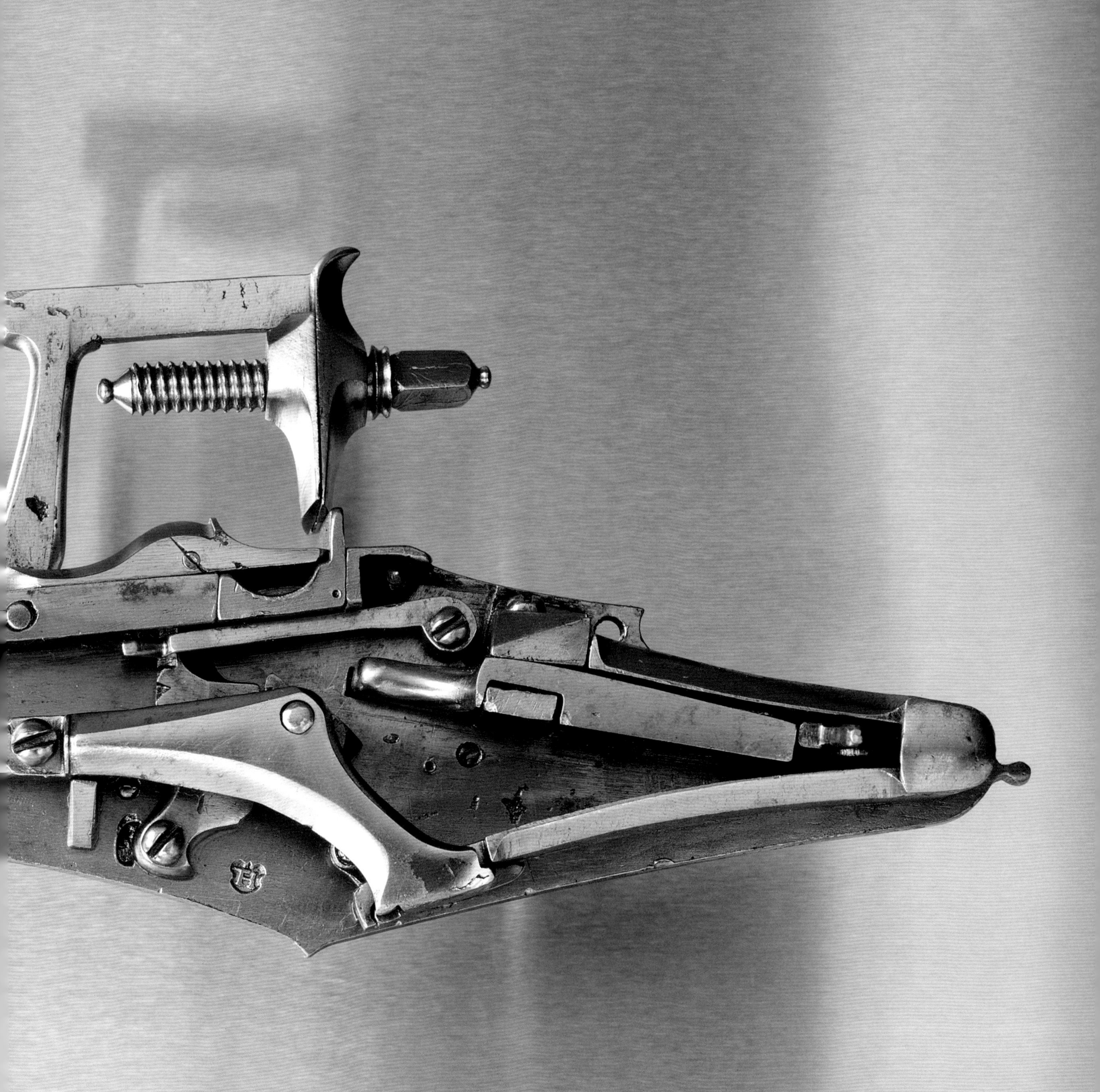

Hungarian Unrest

Whereas the front line was marked by one disaster after the other, Emperor Rudolf in Prague could claim - or so it seemed at the beginning - a series of diplomatic successes. Transylvania, which had been under Turkish rule since the triple division of Hungary, once again came under imperial influence in agreement with its sovereign, Prince Sigmund Báthory. The Transylvanian estates, however, refused to relinquish their rights. Alarmed by Emperor Rudolf's plans to force their dependency as well as religious conversion, they elected the magnate Stephan Bocskay as their new leader. Prince Báthory decided to extend his power beyond Transylvania and entered the war supported by the Turks and an army of "Haiducken" - Hungarian mercenaries - and Tartars.

Among other things, the "Haiducken" captured and looted Fürstenfeld; the year 1605 was characterized by repeated raids and devastation. An inquiry commission listed 3,513 persons murdered or abducted, 1551 houses destroyed by fire and 1,000 animals lost during this year in Eastern Styria.

Finally the imperial general Tilly succeeded in winning control and forced the conflicting parties to accept peace. Bocskay renounced his claim to the Hungarian throne; in exchange the estates retained their political independence and religious freedom.

Emperor Rudolf refused to sign this treaty which had been negotiated by his brother, Matthias. For the latter, it was a welcome occasion to openly carry out his conflict with the Emperor. In this situation Rudolf assured himself of the loyalty of the Bohemians by granting them extensive written concessions. And the Turks, threatened by successful Persian attacks, were forced to accept the 1606 Peace of Zsitvatorok. Although the Turks were accorded the fortress of Kanisza, they lost the right to demand tribute payment, an important privilege dating from the time of Ferdinand I. Yearly the Emperor had been obliged to make the Sultan an "honorable gift" as tangible sign of his subordination. Now the Sultan had to accept the Emperor's equal status. A symbolic defeat of no little importance which strengthened the Emperor's position.

A strengthened position which Rudolf, however, did not exploit further politically. More and more isolated in the "Hradschin" (castle of Prague) by his utopian views of universal rulership, working on scientific as well as alchemical experiments, the Emperor lost hold of real political power and influence. Melancholic and distracted, at the same time arrogant and indecisive, he maneuvered himself into a situation whereby Matthias marched into Prague, had himself crowned King of Bohemia and de facto displaced his brother. After Rudolf's death a year later (1612) the rising differences between the Protestant estates of Bohemia and the Habsburg Counter-Reformation came to an open split. This led after Matthias's death in 1619 to the election of the Protestant Frederick, the Elector Palatine, as King of Bohemia and thus directly to the Thirty Years' War.

For the history of the development of firearms see:
Peter Krenn (editor), Von alten Handfeuerwaffen; Graz 1989.
About the court of Rudolf II in Prague see:
Prag um 1600 (catalog), Essen 1988.
About Giordano Bruno see:
Francis Yates, *Giordano Bruno and the Hermetic Tradition;* London 1964.
Quotations from Michel Foucault are taken from:
Michel Foucault, Vom Licht des Krieges zur Geburt der Geschichte; Berlin 1986.
About questions of taxation and the arms boom on the military border see:
Franz Pichler, Die steuerliche Belastung der Landbevölkerung durch die Landesdefension;
Peter Krenn, Der Rüstungsboom in der Steiermark von 1576 bis 1590;
Karl Kaser, Die österreiche Militärgrenze in Kroatien; all B and B 86.
Information regarding the Counter-Reformation in Styria is based on:
Reiner Puschnig, Ein neuer Adel am Hof und in Innerösterreich;
Karl Kaser, Steirer als Heerführer und Grenzverteidiger; both B and B 86.
Quotations about the salt crisis and border mismanagement are taken from:
Helfried Valentinitsch, Türkennot und Kriegsgewinn; B and B 86.
Details about the course of the "long Turkish war" from:
Günter Cerwinka, Der "Lange Türkenkrieg" und die Haiduckeneinfälle B and B 86

The War is Won

May, 1998

Left over.
Hulls, remnants, war fossils.
Repaired, restored, preserved.
Sent on tour,
USA, Canada, Australia,
and always more along than just decor.
Would the text be different,
had it been written here ...
locked in at night
alone among the weapons...

Exactly a year ago,
for just one night,
as fireworks went off above the Landhaus,
four singers stood here,
in opera costumes, dressed as "Landsknecht" and Turks.
The concert was over,
but still they walked, the opera figures,
together through the house.
Steel theater ...

Halberds/
Cheval-de-frise

War of Races

The old belief, derived from ancient Roman history, that genealogy determines rulership and thus individual sovereignty is largely discarded after Emperor Rudolf's death. Rudolf, who still dreams of himself as being a ruler of divine authority, is removed from all daily political discussion by his utopian views of universal harmony and knowledge and does not act with the times. He overlooks criticism, misinterprets the reactions of others and lacks authority. His fantasy of a genealogical, mystical omnipotence is subsequently replaced by a historical philosophy of victory, defeat and violence.

Michel Foucault calls this new interpretation "biblical" rather than "Roman". The French social historian does not refer to Austrian examples, however had he been familiar with the scenes painted on the ceiling of Eggenberg Castle, commissioned at the end of the 17th century by the Styrian family who had profited enormously from the war, he could not have found a better example for the change in thinking. Inspired by popular historical works of the day, the mixture of biblical and Roman historical narration displaying war, destruction and a new moral postulate is a striking illustration of the new philosophy. "What the one side considers rightful, what represents law and duty to the side of authority, is now shown by this new mode of expression as abuse, malpractice, violence and oppression - as soon as one crosses to the other side ... One now sees rightfulness in its two-faced reality; the triumph of one is the subjugation of the other ... Not only does this "counter-narration" disperse the idea of a unity of sovereign rights and obligations, it also interrupts the continuity of glorification. It makes clear that the blinding light of power does not solidify and fossilize the entire society in a state of constant well-being; on the contrary, the light divides, illuminating only one side of society while leaving other parts in the shadows or pushing them into darkness ... This discursive practice does not carry on the tradition of a long, uninterrupted jurisdiction of age-old powers; it is much more a prophetic rupture. That is why the new discourse also resembles certain epical or mythical or religious forms, which instead of recounting the flawless and unbroken glory of the sovereign, tell of the misfortune of predecessors, of their expulsion and subjugation ... As such this new discourse of the war of races indicates the development of something which is much closer to the mythical-religious history of the Jews than to the political-legendary history of the Romans," concludes Foucault.

Halberds

Two-handed Swords
and Shields

The history of war as a conflict of races (not yet in the sense of "racism" which began to characterize the later 19th century, though heading in this direction) is an attitude which increasingly influenced western assessment of the Turkish wars. The exaggerated prints, handbills and lampoons about *"the Turks"* or *"the Hungarians"* made growing use of simple national associations. The Ottomans are no longer the unknown "others", defined as devilish or mythical hostages of a corrupted Christianity, but people of a certain race. These types of people are associated with characteristics such as "lasciviousness" or "cruelty" and the categorization is verified by citing (horrible) examples.

The so-called "Turk Newspapers" printed news and descriptions of atrocities in song-form. One example is the paper which appeared in 1592 in Graz: "New information and true story...about the hereditary foe, the Turks, who dealt so wretchedly with the poor folk on the Hungarian border to Croatia, all womenfolk and children deplorably murdered and killed ..." The Turkish wars provided a fitting occasion for newly defining history, this time in terms of defeat, murder and the resulting national protest .

New Hungarian Conflict

Emperor Rudolf II's political hesitancy and autism led to the empowerment of Matthias and the famous "fraternal strife in the House of Habsburg". After Rudolf's death, Matthias was crowned Emperor.

The Styrian estates remained neutral and managed to resist the considerable pressure asserted by their own sovereign prince. There has been much speculation about this hesitant attitude - and also as to why the Protestant estates did not openly revolt against Ferdinand, who - elected Emperor in 1619 - was often in political difficulties and showed considerable weakness. Massive resistance could have possibly changed the course of history ... these speculations are useless, however, since all parties concerned were obviously in a state of shock with regard to contemporary developments that must have appeared truly horrible even to hardened cynics. It came to the burning of books, killings, banishment, hypocritical conversions - albeit not surpassing the unparalleled atrocities of the Thirty Years' War.

Edged Weapons and
Round Shields

The most threatening development for Styria during this war did not come from the main military combat, from which the country was largely spared, but instead from a special conflict caused by rebellious Hungarians. Anticipating unstable times, the estates once again boosted their arms supplies around the turn of the century and the armament boom continued undimished. In 1629 an inventory of the province listed 85,000 weapons and pieces of armament; the storage conditions became critical. In 1632 the travel writer Martin Zeiller reported from Graz: " The honorable estates of the province have indeed filled their Zeughaus with a considerable quantity of large equipment, armor and munition; it is only unfortunate that all is so cramped and jumbled."

Though stored in cramped quarters, the equipment was nonetheless often in use since the ambitious prince of Transylvania, Bethlen Gabor, coveted new plans for Hungarian domination. Taking advantage of the confusion in Bohemia, he planned to unite a new Hungarian empire under his leadership and proceeded to assemble a strong, offensive army. Bethlen allied himself with the Protestant Bohemians and from Pressburg led his attack on Vienna. Under these circumstances, the largely Protestant Styrian estates showed their solidarity with the Emperor (a thankless effort), and sent troops to secure the provincial border, though they did not march to defend Vienna. The Emperor's position was not only aided by the "armed neutrality" (Helfried Valentinitsch) of the

Halberds

Styrian estates, but in addition, by the Cossack attack on Transylvania, which forced Bethlen Gabor to retreat. Once the situation there had settled down and he was crowned King of Hungary, military combat flared up again. Bethlen Gabor joined forces with the Turks - who still had problems with Persian invasions and even lost Baghdad - and renewed his attacks. The Inner-Austrian estates reacted by enlisting mercenaries and new troops, whereby a mutiny among the Styrian troops in 1621 caused their dismissal. The war fluctuated between broken peace treaties and repeated attacks until Bethlen Gabor's death in 1629. The arms industry, however, continued to flourish and a new armory for the estates became an urgent necessity.

War Economy

Whereas the Thirty Years' War seriously weakened the Empire, the Ottomans were kept so occupied elsewhere that they were unable to participate in campaigns against the West. Poles, Venetians and above all Persians attacked on various fronts and implicated the Turks in menacing wars so that Emperor Ferdinand II had little trouble in securing further peace settlements from the Turkish Sultan. In addition, internal conflicts weakened the Ottoman Empire; Sultan Osman II was murdered by members of the *Janissaries,* his own elite troop. After the death of Bethlen Gabor the military frontier remained calm, and except for the year 1645

Blades of Estocs and Dusäggen

when Swedish troops devastated a Styrian army, the great battle campaigns and mercenary movements bypassed Styria.
This period of relative stability (despite the raging Counter-Reformation) explains the lively building activity which also included construction of the Landeszeughaus in Graz. However, even as this prestigious symbol of the estates was being erected it representated more figurative than actual power. The pride of the landed lords had been broken by the Counter-Reformation; the Provincial Diet was reduced to little more than an instrument for tax allocation, completely dependent on the reigning prince - to whom many newly ennobled aristocrats were indebted, not to mention the monasteries and cloisters which enormously benefited from their new security and now made exorbitant profits.
Instead of honoring the old traditional social order, the old/new landed lords strived to increase their earnings. The feudal relationship of responsibility becomes - thus concluding a development over the last 150 years which is repeatedly mentioned in the previous chapters - a relationship of pure subjugation. The nobility deferred to the reigning prince while demanding subjugation from the subjects.

Compulsory work became real forced labor, more and more to the exclusive benefit of the landowners, and led to mass impoverishment of the peasantry.

Conditions among the population continued to worsen. Though most endangered regions in Styria were spared military conflict, the economic situation was catastrophic. Old trade routes and economic relations were disrupted by the confusion of war, inflation threatened the monetary system (in 1623 the imperial government had to declare bankruptcy) and only a few businesses made profits. As for instance, the cattle-dealer from Pettau, Matthias Gualandro, and the grain and cattle-dealer from Wildon, Hans Edelman, who both made fortunes during this period while others headed toward financial ruin.

Coupled with inflation was a raise in food costs, finally a food shortage and ensuing famine. Besides climbing costs, taxation increased constantly - meanwhile approved without contradiction by the Provincial Diet. In 1632 a personal tax was imposed on every inhabitant of the province. In 1635 a tax on capital and capital interest followed, and finally in 1640 a chimney tax placed an additional duty on every hearth. In 1642 the Pro-

Morning Star | l 236,4 cm
Egid Rotter
Graz, 1685

Military Fork | 214,2 cm
17th century

vincial Diet devised a property tax to be levied on all remaining property, and in 1645 a revision of this property tax was introduced along with obligations which placed a further toll on public officials.

Apart from the increasingly oppressive taxation, forced labor and inflation, another immediate danger - this time from outside - threatened the rural population. Styria was the assembly point for imperial mercenaries who took up quarters in cities and villages and at times reigned by terror. Murder, rape and looting did not reach the same dimensions as in the immediate combat zones, however, they were not infrequent occurrences. At first this provoked only isolated attacks on soldiers by armed farmers; finally in 1635 it came to an organized peasant revolt on the military border. This uprising was brutally crushed, but the remaining discontent and sheer misery caused defiance and rebellion to flare up time and again.

Height of the Turkish Wars and Victory

Following the major war in central Europe which subsequently changed the balance of power, a system of absolute sovereignty grew out of the chaos. Similarly, shifting spheres of influence in the Ottoman Empire aided in strengthening the weakened central authority. Along with the sultans, the grandviziers of the Köprülü family established themselves as a guaranty of political stability - with their own military ambitions. The final important conflict between the Ottomans and the Empire broke out once more in Hungary.

Since 1657 Leopold I had reigned as Emperor in Vienna. His renewal of the Habsburg claim to Transylvania provided the Turks with sufficient reason to mass troops. In 1663, accompanied by an army of 100,000 men,

Nail-studded Flail | 238 cm
17th century

the Grandvizier Achmed Köprülü captured the fort of Neuhäusel in a surprise attack. With the fall of this last major bastion on the way to Vienna, the Emperor's position became precarious. While the Turks were preparing their winter quarters, the Emperor quickly solicited allies and organized a large defense army. Under the supreme command of Imperial Field Marshal Raimund Count Montecuccoli a coalition army of imperial, French, Italian, Spanish, Swedish and German troops was assembled (whereby the customary looting and ransacking of native territory had to be endured).

The Styrians feared the worst and once again prepared accordingly, laying in food supplies and assembling local troops. They also made a tactical mistake with fatal consequences. In 1664, against the wishes of Montecuccoli, Styrian troops attacked the fortress of Kanisza which was in Turkish hands and thus considered an immediate threat. However, instead of capturing the fort, they only succeeded in driving the central Ottoman assault toward Kanisza. The Styrians were forced to withdraw with tremendous losses and Montecuccoli's tactical plan was ruined. Nonetheless the outstanding strategist succeeded in dealing Turkish troops the

Military Scythe | 226 cm
17th century

decisive blow. At the Battle of Mogersdorf imperial forces were victorious. A new peace treaty was quickly concluded with the Sultan. Realistic, if one considered the tremendous consequences of this bloodbath for both sides; ridiculed, however, as the "shameful peace of Eisenburg" since one had hoped after this first major victory over the Turks, to finally overpower them for good.

First, however, it came to a further existential crisis. In 1683 the Grandvizier Kara Mustafa, inspired by dreams of world power, attacked and overran the border fortifications with an army of over 200,000 men, and then pushed forward to Vienna, spreading panic and horror along the way. The city was saved by Pope Innocent XI's diplomatic idea of a "Holy Crusade" against the Turkish threat, which managed to ally such contrary leaders as the Polish King Jan Sobieski, the Venetian Dukes and the Emperor. This relief army, led by the Polish King and Duke Charles V of Lorraine, won a resounding victory at the battle of Vienna, the Turks fled - Sultan Mohammed IV had the unfortunate Kara Mustapha killed by strangulation - and the counter-offensive pushed far into Hungary. Further successful campaigns under the leadership of Prince Eugene of Savoy secured the Emperor's position

Bill | 230,6 cm
17th century

and in 1699 brought Hungary and Transylvania back under Habsburg control. After this the Ottoman Empire was never again to present a serious threat - endangerment from the Turks was over. In 1699, the same year that Hungary was reclaimed, a new inventory of the Zeughaus in Graz recorded the largest number of weapons and armaments ever to be stored there, a total of 185,700 pieces. This record level simultaneously marked the beginning of the decline into obsoleteness. The military frontier was no longer necessary - minor skirmishes which occurred in the area during the next 50 years were insignificant compared to the Turkish wars - and the centralization of military authority in Vienna made the outfitting of occasional troops less and less frequent. Once again, during another Hungarian rebellion between 1703 and 1711 (the so-called "Kuruzzen Wars") there was plundering and devastation in Styria, which was then followed by peace. The situation of the peasant population, however, did not improve. Although spared of foreign attacks, they were still subject to exploitation by the landed lords; the era of lawlessness continued.

Military Sickle | 226 cm
17th century

Discipline

At the beginning of the 17th century, Montgommery wrote in his book *The Art of War:* "Furthermore, one must accustom the soldiers to marching, as it is called, in rows, both when covering territory or in battalion, so that they march to the beat of the drums. And to make this better possible, they should commence with the left and end with the right foot, so that together the entire troop lifts one foot and puts the other down at the same time." A good hundred years later, in 1766, the directions for marching sound as follows: "The length of the short step is one foot, of the normal, the double and the street step two feet, measured always from heel to heel. As for the allowable time, one has a second for the short step and normal step, half a second for the double step and a little more than a second for the street step. The diagonal step also takes one second; it is no more than 18 inches long. One takes the normal step forward, with raised head and erect posture, balancing alternately on one leg while

Mail Shirt Early 16th century

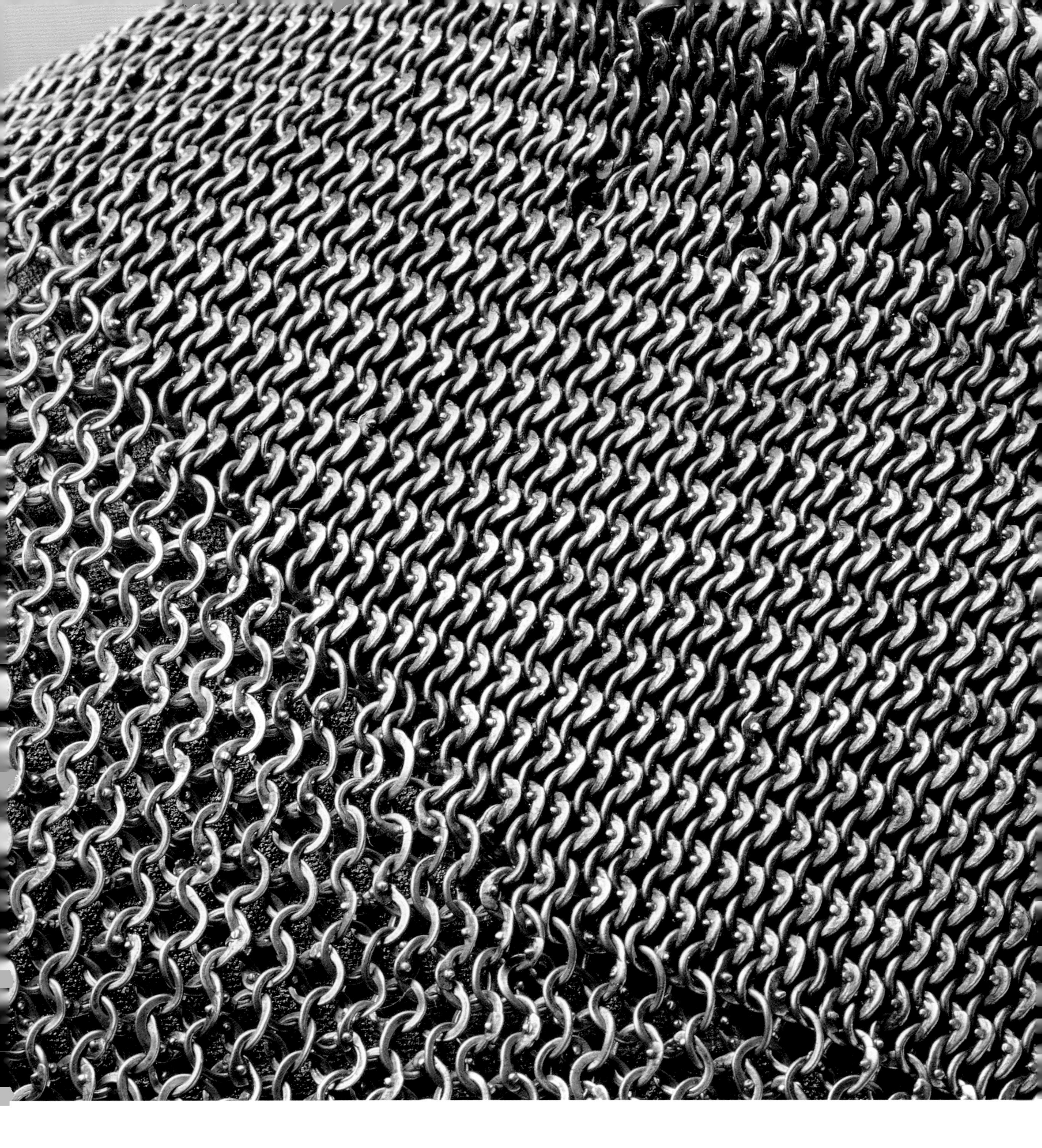

lifting the other forward; the knee is taut, the top of the foot is pointed slightly outwards and bent down so as to lightly brush the ground on which one marches, placing the foot in such a manner that all parts touch down simultaneously, without kicking the ground. "The structures of time, the human body and society - all are meanwhile influenced by mechanization, which functions by regimenting the body. Discipline implies a submission of the body, but it also regulates thought.

Mechanisms and structures set the standards to which humans must adapt. And the language of the military can be taken pars pro toto as the society's ideal. Just as the soldier after the mid-18th century becomes a norm, so are the basic social relations and structures adapted to him. Just as the machinery of new manufacturers dictates the pace of work, so does the possible speed of recharging weapons determine the tempo of war. Just as the seconds, minutes and hours of a day are divided into standardized rhythms, so are people's lives determined by this daily plan. When and what has to be done, who is in which position ...

Halberds

First half of the 16th century

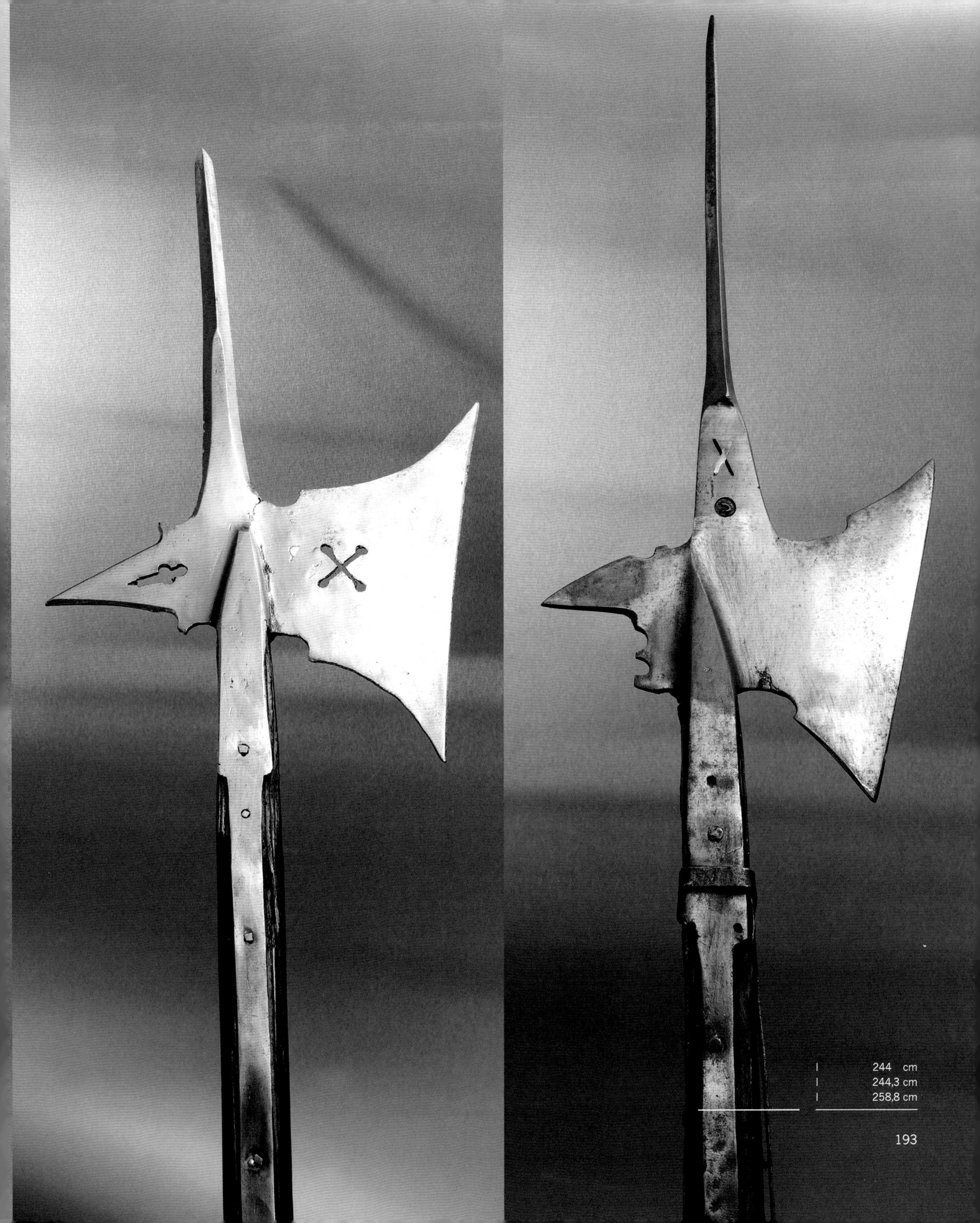

| 244 cm
| 244,3 cm
| 258,8 cm

Historicism

Under these circumstances, recollections of Zeughaus days are indeed mainly nostalgic. And in 1749 one began to remove old furnishings and place the weapons in decorative arrangements. Pyramids and columns were used with draperies, names of weapons and arms were spelled out in pistols. In an age when the military functions through discipline, the Zeughaus equipment, which still related to individual needs, appears as an oddity. Legends grew up around the weapons - this is the beginning of chivalrous romanticism which sets in once all true relation to the arms and equipment is lost. Nothing is so un-modern as the immediate past, and thus an early historicism takes hold of the Zeughaus.

One may also interpret history as follows: By the time the Styrian estates requested permission to preserve the armaments (which Maria Theresia's imperial court considered "scrap metal"), as a reminder of their prestigious past, this prestige was no longer part of the political reality. The symbols of former power become decorative elements in a fading recollection, detached from concrete incidents. As soon as the Turks no longer present a threat, they become popular figures and the tales of war become legends. In his *Abduction from the Seraglio* Wolfgang Amadeus Mozart evokes no more than a slightly uncomfortable shudder when the

Halberd

First half of the 16th century
l 258,8 cm

harem guard threatens to decapitate the Christians and "impale them on hot rods". The protagonist of the opera, Bassa Selim, is now much more than a "noble savage", he is an enlightened monarch who speaks of understanding as well as of political power. The swords and spears, arms and pistols of the period of the Turkish Wars are also seen in a new context. As "pretty-gruesome" remnants of a "graying past" (as Karl Meyer calls them in his "Versuch über Steyermärkische Altertümer" published in 1782) which, since they appear obsolete, may now be enjoyed.

The Zeughaus has a national, popular character - yet it symbolizes something which has been lost. Warfare has changed, society as well. Again Michel Foucault: "It may be that war as strategy is a continuation of politics. But it must not be forgotten that "politics" has been conceived as a continuation, if not exactly and directly of war, at least of the military model as a fundamental means of preventing civil disorder. Politics, as a technique of internal peace and order, sought to implement the mechanism of the perfect army, of the disciplined mass, of the docile, useful troop, of the regiment in camp and in the field, on maneuvers and on exercises. In the great eighteenth-century states, the army guaranteed civil peace no doubt because it was a real force, an ever-threatening sword, but also because it was a technique and a body of knowledge that could project their schema over the social body. It there is a politics-war series that passes through strategy, there is an army-politics series that passes through tactics. It is strategy that makes it possible to understand warfare as a way of conducting politics between states; it is tactics that makes it possible to understand the army as a principle for maintaining the absence of warfare in civil society."

What now remains for the Zeughaus of Graz is historicism. One day other generations will judge our times, our conception of "reconstruction" and "original", the efforts to preserve and re-establish what must have been once "authentic". And surely they will learn as much about us as about the decisive times of others.

Quotations concerning the "war of races" are from:
Michel Foucault, Vom Licht des Krieges zur Geburt der Geschichte; Berlin 1986.
I have taken facts and quotes about the "Turk Newspapers" from:
Theodor Graff, Kriegsberichte - Türkendrucke aus Graz; B and B 86.
Dates and descriptions of Zeughaus construction from:
Peter Krenn, Das Steiermärkische Landeszeughaus in Graz; Graz 1974.
Descriptions of historical events during the Turkish and Hungarian Wars according to:
Günter Cerwinka, Der "Lange Türkenkrieg";
Helfried Valentinitsch, Die Bedrohung der Steiermark durch Bethlen Gabor,
Helfried Valentinitsch, Der dreißigjährige Krieg;
Josef Riegler, Höhepunkt und Ende der Türkengefahr; all B and B 86.
The quotations concerning military discipline from:
Michel Foucault, *Discipline and Punish,* translated by Alan Sheridan, 1977.

Halberd

Late 16th century
l 213,6 cm

Halberd
for Guardsman

Pankraz Taller
1575–1600
l 228,4 cm

Glaive
1575–1600
l 257,7 cm

Blade with Mark of
Peter Schreckeisen

Epilogue

Brief Remarks Concerning Military Technology and Weapons

A history of the Zeughaus of Graz must always include a history of weapons. Yet this history of weapons must similarly be accompanied by the thought of injury and death. A study of arms and armaments is almost impossible without considering what they intended to achieve or prevent. Historical descriptions of weapons are filled with horror and fascination. In some older arms books the text reads like the equivalent of a firearms licence. By contrast, others deal with the subject in a remote scientific manner, formally dissecting and classifying the objects as if to conceal their real function. Is a "politically correct" approach to weapons possible - or even desirable? To deny the strange attraction which the power of weapons exert would be to ignore an important element out of a false

Friuli Spear | Late 16th century | l 273,6 cm

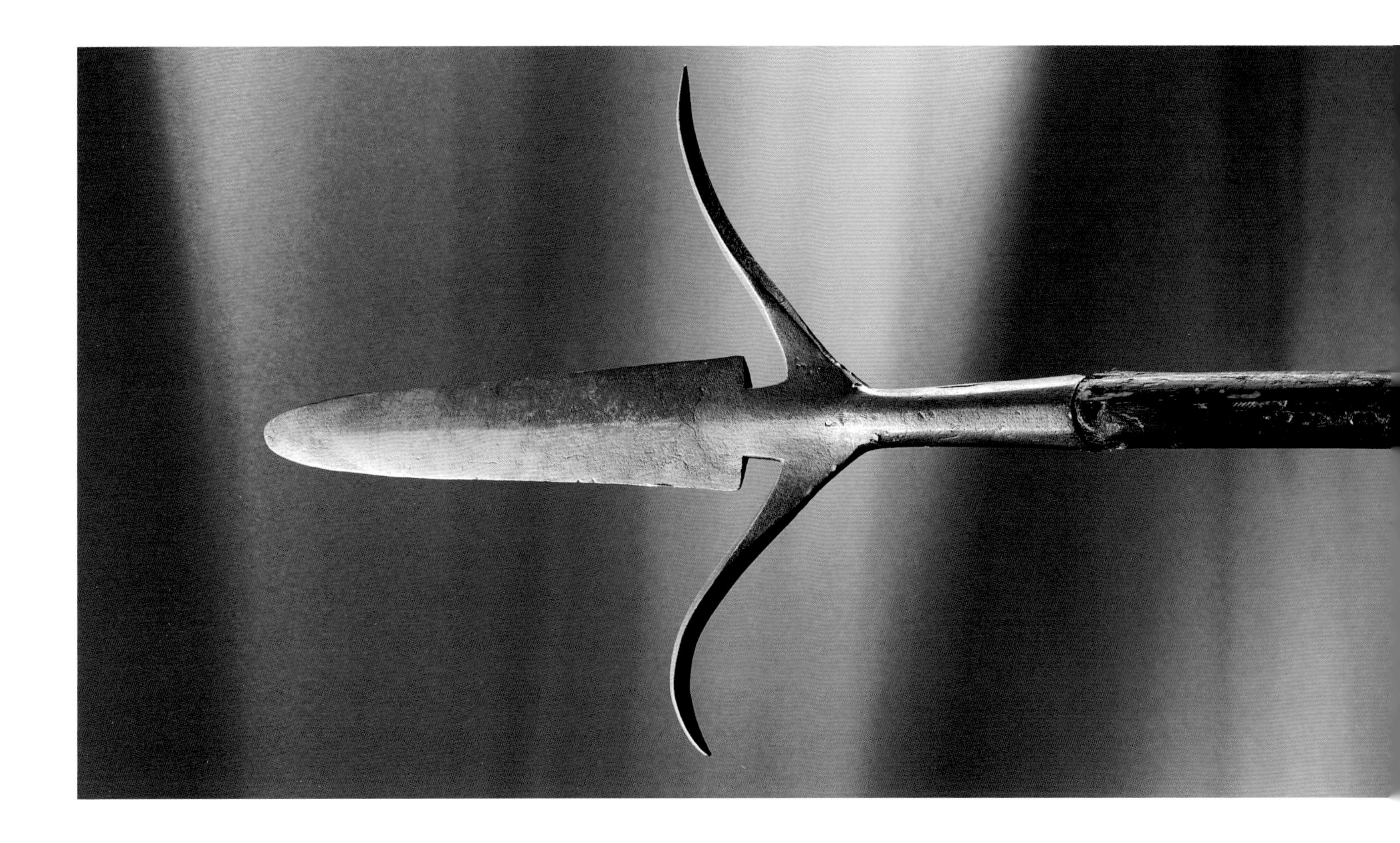

sense of shame. After all, it must be stressed that reactions to weapons are often paradoxical; conflicting feelings should not be subjected to rash moralizing but rather to careful analysis - including one's own motives for dealing with the subject. That weapons exist in reality, that they were and still are employed today is certainly an important factor. The following glossary provides a brief functional and historical classification of weapons. Their technical development and the consequences for warfare in general are illuminated in the epilogue.

Velocity

Under the impression of the Golf War, which news broadcasting stations like CNN portrayed as a virtual video game without victims, Paul Virilio also applied his theory of the alienation resulting from acceleration to the case of war. Indeed, the importance of distance, as discussed in previous chapters with regard to the Zeughaus weapons, only makes sense in connection with acceleration. The speed of military movements is

Spetum — Late 16th century — l 213.5 cm

decisive in winning battles. Even ancient troops under Julius Caesar achieved their superiority by conducting colossal marches at a tempo considered almost impossible for the day. Caesar's troops were quick, their sudden appearance surprising.

Distance and velocity coincide with the development of long-range weapons. Just as crossbows and longbows were responsible for the decline of medieval knightly troops, firearms multiply the importance of the relationship between speed and distance. Firearms wound or kill at a longer range. Destruction is either widespread (heavy artillery) or precise (hand firearms). New ignition mechanisms are mainly responsible for the rapid improvement in weapon technology. Match-, wheel-, and finally flintlock weapons, each more reliable than their predecessor, are combined with possibilities for more rapid loading and firing. The faster a weapon can be loaded, the more effective it is. The Prussian army of the High Baroque era was above all successful by consequently drilling its soldiers to achieve a rapid, collective acceleration of the weapon-loading process. A German proverb, which aims at self-reassurance by claiming that "the Prussians don't shoot so fast", testifies to just this fact: superiority due to speed.
The development of war technology is thus based on the components of flexibility (variability of troop movements with fast marching footsoldiers, who build "human walls" during battle to hinder the enemy, and a cavalry capable of rapid, surprise attacks) , distance (increased range of weapons and, after the replacement of stone balls by explosives, also a new lethal effectiveness) and velocity.

Throughout the history of weapons, velocity has been increased in a variety of ways. Whereas rapid weapon-loading during the High Baroque period was based on drill and required large numbers of qualified troops, the addition of magazines to rifles later sufficed to enable constant firing. The modern machine-gun transfers acceleration to the weapon itself. Lethal effectiveness is improved by longer range, greater penetrating power and increased shooting frequency. All criteria which had proven militarily effective in early firearms and which - on a more primitive scale - had

Spear 1575–1600

l 265,3 cm

Ceremonial Partisan for the Guard of Franz von Stainach		1628
	l	249,4 cm
	w	3,02 kg

been used during Zeughaus days: By improving the ignition mechanism and shortening the barrel it was possible to equip the cavalry with firearms. Pistols and rifles were no longer immobile. (Earlier rifles were so heavy that they had to be firmly anchored to absorb the powerful kickback. The large Landsknecht rifles also had a supporting stand to increase their accuracy, thus making them quite unwieldy and immobile.) Riders with small firearms combined the speed of mounted attacks with the longer range of firearms plus the increased frequency of shots.

In the face of this increased velocity armament soon became a hindrance. It reduced the mobility of troops and could no longer withstand the penetrating power of the projectiles.

The speed of bullets required such a proportional increase in the thickness of the armor, that a person could no longer move in it. (Not until the use of armored vehicles during World War I were armor and mobility reunited.) Body armor which is too thin and thus penetrable, can even in-

Iwo-handed Sword
Blade 14th century
Hilt 16th century

Passau
l 169 cm
w 3,76 kg

crease the degree of injury since the bullet, slowed by its impact, lodges in the body, possibly splintering and causing greater tissue damage.

The distance between victim and weapon-user in modern times has taken on sheer universal dimensions as seen by the development of air force, rocket technology and finally intercontinental missiles. Even the ballistics for canons in their heyday required complex mathematical calculations. During World War II the distance between gunners and targets lay so far apart that determining the projectile trajectory took several days and the joint effort of dozens of people. In the American Army these trajectory calculators were called "computers". Their interest in the mechanization of this task led to the first large scale calculator, which was subsequently employed for military purposes and called a "computer". It now unifies the parameters of velocity, distance and flexibility in an entirely new dimension. The Zeughaus of Graz represents a significant stage in this historical development. As such its past leads as well to the present.

Two-handed Sword | Styria | 1575–1600
l | 171 cm
w | 3,51 kg

Hungarian Saber 1575–1600

Broadsword 1675–1700

Glossary

ARMORER

Craftsman who produced armor; divided into mail and plate armor makers.

ARQUEBUS (harquebus)

Short gun; weapon used by the arquebus-riders, lightly armed riders who came into use in the second half of the 16th century.

ARQUEBUSIER

Footsoldier equipped with light gun.

BANDOLEER

A leather strap worn across the body by musketeers for carrying powder flasks and ball pouches.

"BESCHAU" or view-mark

Craftsmen marked the weapons and armor they produced with individual stamps. In addition, in large centers of armor production, such as Nuremberg or Augsburg, city examiners made a final inspection of the armor in order to insure the city's reputation. This stamp of approval, the "Beschau" or view-mark, was thus an important guarantee of quality.

BILL

A weapon used up to the 18th century that consists of a long staff terminating in a hookshaped blade, used to sever the fetlocks of a horse.

BULLETS

Ammunition for firearms. Until the 19th century, bullets used in firearms were mostly made of lead or lead-coated iron. Two types of bullets must be distinguished: ball bullets which rolled easily in the barrel (were quickly loaded but not as accurate) and cartridges which were rammed down into the barrel with a ramrod. This was more accurate but also more time consuming.

CHEVAL-DE-FRISE

Wooden beam through which short pikes (such as those developed for boar-hunting) are stuck in a crosswise alignment. Chevaux-de-frise were used at the end of the 17th century as obstacles to prevent enemy approaches and to protect the infantry behind them from cavalry charges.

CROSSBOW

Long-range weapon, probably originating in China and described for the first time as "gastraphetes" by the Greeks in the 3rd century B.C. Bow with strong string mounted on a middle shaft. The medieval crossbow was readied for firing by means of a mechanical device (lock system with trigger: the string rests in a notch, bolt or quarrel is mounted in place, pressure on the trigger releases the string). The crossbow is highly powerful but sensitive to moisture and less accurate than the longbow.

CUIRASS

From the French "cuirasse", originally "leather armor"; combination of breast- and backplates worn by cuirassier.

CUIRASSIER

A heavy cavalryman.

CUISSE

Upper part of the leg harness, protecting thigh and knee.

DUSÄGGE

A short infantry saber; name derived from the Czech word "tesak" meaning "long knife". The dusägge combines an oriental curved blade with an elaborated European-style hilt (both handle and protection for the hand).

EDGED WEAPONS

General term to classify weapons for cutting, thrusting and dealing blows.

ESTOC

A light Renaissance version of the thrust sword of the Late Middle Ages.

FIELD ARMOR

Complete suit of armor for heavy cavalry.

FIELD HELMET

Helmet with visor and neck-lames that came into use between 1520 and 1530.

FIREARMS

Long-range weapons fired with gunpowder. Consisting of salpetre, sulphur and charcoal, gunpowder was first mentioned in Europe around the mid -13th century by the Friar Roger Bacon, who also published the formula for its production. Firearms using gunpowder were first used in Italy in 1326; their lethal effectiveness made them spread rapidly. Firearms are divided into two groups: large artillery and hand firearms.

FLINTLOCK

A further development of the snap-lock, the flintlock was invented around 1610 in Normandy, and for 200 years served as the most important ignition system for hand firearms. The flintlock is based on the same principle as the snap-lock, but in a more perfected form: pressure on the trigger causes the engaged cock with flint to snap forward, the flint sharply strikes the combined pan-cover and steel. Hitting the vertical face of the steel, the flint produces sparks, while at the same time the force of the cock impact pushes the pan-cover and steel away, exposing the priming powder to the sparks and thus igniting it.

FLUTING

Raised metal ribbing imitating folds of material; used as decoration for German armor of the early 16th century.

GLAIVE

A hafted weapon with a long, knife-like head fitted in line with the shaft, used by Habsburg guards in the 16th century.

GREAVE

Armor for the leg below the knee.

GRÜNDERZEIT

Era of reckless financial speculation following the Franco-Prussian War, 1871-1874.

GUNPOWDER

Used as propellant, as well as means of ignition in firearms; made of saltpeter, sulfur and charcoal. Since the mid 15th century produced in large quantities in powder mills, whereby the saltpeter dissolved in liquid acts to bind the other ingredients. The mixture is dried and pounded into a cake, then broken into small bits and run through a sieve. The "grained powder" is stored in "powder flasks" and used as propellant being rammed into the barrel of the weapon. A ball-bullet or other projectile is then rolled in or loaded onto this bed. Finely ground powder (priming powder) on the touch-hole which also leads to the main charge is ignited by various mechanisms (see types of locks) and the powder in the barrel explodes discharging the bullet.

HACKENBÜCHSE

German name for heavy iron firearms (literally "hook-gun") in use since the 15th century. The hooks served as a recoil-absorbing device by anchoring the weapon to walls or beams.

HALBERD

Name derived from the German words "Halm" (staff) and "Barte" (ax). The long staff weapon combines the sharp spike of a spear with the blade of a battle-ax and a hook-like fluke. It was especially useful in combat against heavily armed riders. After the triumph of firearms and subsequent disappearance of knightly troops, the ax function became useless and halberds were increasingly replaced by long spears.

HARNESS

A general term for body armor worn both in combat or during sporting tournaments. Also for horses (full bard). Made of stiff metal plating or flexible metal rings which are riveted together.

HAUBERK

The medieval French name for a long shirt of mail extending to below the knee. The hauberk was then replaced by a shorter version, the haubergeon.

HELMET

Common name for various forms of armored head gear. Either as "close helmet" entirely encasing the head (with visor) or as "burgonet" with peak or fall over the brow, otherwise open. Numerous variations for special purposes.

HISTORICISM

A theory that emphasizes the importance of history as a standard of value.

HUSSARS

Originally lightly armed riders from Hungary (Matthias Corvinus called them "hussarones"), from the 16th century also part of the imperial army; outfitted with sabers, estocs, maces and spears and protected by a mail coat, cuirass and the Hungarian-style helmet (Zischägge).

MAIL

An ancient type of body armor comprised of a mesh of interlocking, riveted metal rings. Capes of mail and mail shirts were used as knightly armor well into the 12th century.

MARKSMAN'S HELMET

Helmet with small or moveable brim enabling marksmen to position and aim firearms without hindrance.

MATCHLOCK

The first system of mechanically applying ignition to a firearm. Developed from the 15th century, it solved the problem of holding and aiming the weapon while igniting the gunpowder. The sharpshooter clamps a length of smoldering slow-match in the jaws of a cock or serpentine. Manual pressure on a trigger (later lever trigger) presses the match into the pan of priming powder which then ignites. This means of ignition is very sensitive to moisture since the match smolders openly. The slow-match or fuse was made of hemp cord dipped in a solution of lead acetate. In addition "fire-mushrooms" were cooked, dried and cut into strips as tinder; these mushrooms contained saltpeter and burned well.

MILITARY FORK/MILITARY FLAIL

Rudimentary weapons made by peasants from agricultural tools.

MORION

Helmet used by infantry, with a tall comb and curved brim that sweeps up into pointed ends (also called "marksman's helmet with Spanish form").

MORNING STARS

Stabbing and striking weapon of wood with iron spikes mounted in the striking head. The morning stars in the Zeughaus of Graz date from 1685 and have, in addition to the spikes, a long, square thrust blade attached to the end.

MUSKET

Heavy, long-barreled firearm; used by musketeers, 17th century foot-soldiers

PALLASCH

A type of sword used by Hussar cavalrymen whereby the sword blade is symetrically mounted on a saber hilt.

PARTISAN

The name of this large staff weapon with wide, often decorated blade derives from the Italian word "partigiano" or "fellow party member". The partisan was a popular ceremonial arm for parades and guardsmen; from 1600 worn mostly by officers.

PIKES

Pikes were the most important weapon of the infantry, first used widely by the Swiss against the Habsburgs. Pikes were long (up to 6 meters) thrusting arms with metal blades, used in mass formation to avert cavalry attacks. The pikemen (known as "Knechte" or later "Pikeniere") lined up in several rows one behind the other with their pikes anchored to the ground or held straight forward, and protected the musketeers while they reloaded during cavalry attacks.

PROOF FIRING

The firearms delivered to the Zeughaus of Graz, though produced in series, were still basically individual, handcrafted pieces. Nonetheless, for reasons of security, every piece was tested under the supervision of the Zeughaus armskeeper often to the musical accompaniment of drums and fifes. For this so-called "proof" firing, a larger amount of gunpowder than necessary was often used to test the durability of the barrels. If the weapons stood the test they received the "imperial stamp" and were admitted for storage in the Zeughaus. Similarly, occasional tests to control the material quality of the armor were made by firing at them. Most of the numerous dents visible on the Zeughaus armor resulted from such trials and not from use in combat.

SABER

Old Mongolian cavalry weapon brought to Europe by the Turks; curved, single-edged blade, adopted by the Hungarians and later used by imperial troops.

SNAP-LOCK

Further development of the matchlock and the wheel-lock: the pre-tensed cock spring is released by pushing the trigger. A burning match (match-snaplock) or flint held in the cock jaws is used to ignite the charge. A flint can be used repeatedly, between 30 and 50 times, before it must be changed.

SWORD

Along with the dagger, the sword is the oldest edged weapon with symmetrical structure developed in the early Bronze Age: a straight, double-edged blade placed on a thorn-like prolongation called the tang, held in early days on both ends by two ferrules, between which the grip was clamped. The upper hilt ferrule developed into the later pommel; the lower hilt ferrule was replaced by quillons which protected the hand from blows. During the Renaissance, the lower hilt became a very elaborate hand protection, often with additional side-rings to shield the fingers.

The sword was used as a thrust and blow weapon. One variation is the two-handed sword ("bidenhander"), employed from the 15th century by foot troops against rows of lancers. The long swords - often with double-edged, wavy (flaming) blades and additional small parrying hooks - had to be twirled rapidly through the air in order to clear a path through lines of enemy lancers enabling the cavalry to push forward. Soldiers who mastered the art of the two-handed sword received double pay.

TASSETS

Laminated or one-piece (rigid) extension of the armor skirt for protection of the loins. "Long tassets" were laminated tassets protecting the thighs down to the knees.

VEMBRANCE

Piece of armor protecting the arms.

VISOR

Moveable facial protection of a helmet, may be either raised or lowered on hinges, with slits for vision and ventilation.

WHEEL-LOCK

Mechanical system of ignition for firearms, developed around 1500 in Nuremberg, probably from old friction lighters, whereby a small steel wheel rapidly rotates against a piece of iron pyrite thus producing sparks. The wheel-lock used in firearms functions by the same principle. Pressure on the sear lever releases the pre-set, spring-tensioned wheel. This steel wheel rotates rapidly producing sparks by rubbing against a piece of pyrite held in the jaws of a cock. The gunpowder is ignited by the sparks. Wheel-locks were common into the 17th century. They could be mounted in shorter guns such as arquebus or pistols which were then used by the cavalry. One disadvantage was the fact that the mechanism dirtied quickly - so that matchlocks and wheel-locks were often used in combination.

ZEUG

Old German term for everthing which has to do with ordnance. Zeughaus (armory) is the place where this equipment is housed. Landeszeughaus is the armory of the province.

ZISCHÄGGE

Helmet worn by the Hussars

Exact descriptions of weapons and arms types as well as historical information and research regarding ordnance may be found in the following publications of the Landeszeughaus of Graz:

Peter Krenn (editor), Harnisch und Helm; Ried 1987.

Peter Krenn (editor), Rifles and Pistols; Ried 1990.

Peter Krenn (editor), Swords and Spears; Ried 1997.

Peter Krenn (editor), Von alten Handfeuerwaffen; Graz 1989.

Thomas Höft, born 1961 in Lüchow

Study of art history, literature, language and musical arts

Author in the field of fine arts and music, he is above all well-known for librettos and stage-plays such as "Circe and Odysseus" (Komische Oper Berlin), "Castor and Pollux" (Academy of Arts Berlin) or "Heinrichs Fieber" (Kleist Theater Frankfurt an der Oder).

In 1995 his modern version of Suppé's "Galathee. Die Schöne" was staged by Götz Friedrich in the Deutsche Oper Berlin.

In 1996 he collaborated on the much acclaimed exhibition "Between Heaven and Earth" for the Landesmuseum Joanneum.

Since 1995 as author and dramatic advisor for the "styriarte" festival.

Alexander Kada, born 1968 in Graz

Architectural studies

As international designer in the area of graphics, architecture and interior design, his work includes photo conceptions and numerous book and catalogue projects.

His main activity centers around a broad conception of corporate design for both cultural institutions and large corporations, as well as private clients.

His customers include: Vola International, Toni&Guy Hairdressers, Architectural Foundation of Austria, Europan Austria, Landesmuseum Joanneum, Grazer Kunstverein, Cmst-Power Company, Provincial Government of Styria, House of Architecture.

Angelo Kaunat, born 1959 in Munich

Architectural studies

After several years as lighting and stage technician for the Kammerspiele München, he began working as light designer and stage manager for theater, fashion shows and television productions, later in various architectural offices in Paris, Munich and Graz.

Since 1991 free-lance photographer for prominent technical periodicals (Architecture & Techniques, DBZ Deutsche Bauzeitung, Bauwelt, Werk Bauen Wohnen, Baumeister, Domus, Architectural Review, Architektur Aktuell, Disegno) and for architects.

The theme "light-space-man" dominates his artistic interest.

sappi

Landesmuseum Joanneum

Graz, Austria

Title of the original edition: "Welt aus Eisen. Waffen und Rüstungen aus dem Zeughaus in Graz"

Translated from German by Martha Davis Konrad

Printed in Austria

Text: Thomas Höft

Book design and photo concept: Alexander Kada

Photography and lighting: Angelo Kaunat

Consulting curator: Raimund Bauer

Landeszeughaus

Herrengasse 16, 8010 Graz, Austria

Director: Peter Krenn

Printing: Grasl Druck & Neue Medien, 2540 Bad Vöslau, Austria

Paper: Magnomatt Satin 170 g, acid free and chlorine free, graciously provided and specially produced for this edition by sappi, 8101 Gratkorn, Austria

SPIN: 10664416

With 127 colored figures

ISBN 3-211-83098-7 Springer-Verlag Wien New York